Selected Titles in This Series

39 **Larry C. Grove,** Classical groups and geometric algebra, 2002
38 **Elton P. Hsu,** Stochastic analysis on manifolds, 2001
37 **Hershel M. Farkas and Irwin Kra,** Theta constants, Riemann surfaces and the modular group, 2001
36 **Martin Schechter,** Principles of functional analysis, second edition, 2001
35 **James F. Davis and Paul Kirk,** Lecture notes in algebraic topology, 2001
34 **Sigurdur Helgason,** Differential geometry, Lie groups, and symmetric spaces, 2001
33 **Dmitri Burago, Yuri Burago, and Sergei Ivanov,** A course in metric geometry, 2001
32 **Robert G. Bartle,** A modern theory of integration, 2001
31 **Ralf Korn and Elke Korn,** Option pricing and portfolio optimization: Modern methods of financial mathematics, 2001
30 **J. C. McConnell and J. C. Robson,** Noncommutative Noetherian rings, 2001
29 **Javier Duoandikoetxea,** Fourier analysis, 2001
28 **Liviu I. Nicolaescu,** Notes on Seiberg-Witten theory, 2000
27 **Thierry Aubin,** A course in differential geometry, 2001
26 **Rolf Berndt,** An introduction to symplectic geometry, 2001
25 **Thomas Friedrich,** Dirac operators in Riemannian geometry, 2000
24 **Helmut Koch,** Number theory: Algebraic numbers and functions, 2000
23 **Alberto Candel and Lawrence Conlon,** Foliations I, 2000
22 **Günter R. Krause and Thomas H. Lenagan,** Growth of algebras and Gelfand-Kirillov dimension, 2000
21 **John B. Conway,** A course in operator theory, 2000
20 **Robert E. Gompf and András I. Stipsicz,** 4-manifolds and Kirby calculus, 1999
19 **Lawrence C. Evans,** Partial differential equations, 1998
18 **Winfried Just and Martin Weese,** Discovering modern set theory. II: Set-theoretic tools for every mathematician, 1997
17 **Henryk Iwaniec,** Topics in classical automorphic forms, 1997
16 **Richard V. Kadison and John R. Ringrose,** Fundamentals of the theory of operator algebras. Volume II: Advanced theory, 1997
15 **Richard V. Kadison and John R. Ringrose,** Fundamentals of the theory of operator algebras. Volume I: Elementary theory, 1997
14 **Elliott H. Lieb and Michael Loss,** Analysis, 1997
13 **Paul C. Shields,** The ergodic theory of discrete sample paths, 1996
12 **N. V. Krylov,** Lectures on elliptic and parabolic equations in Hölder spaces, 1996
11 **Jacques Dixmier,** Enveloping algebras, 1996 Printing
10 **Barry Simon,** Representations of finite and compact groups, 1996
9 **Dino Lorenzini,** An invitation to arithmetic geometry, 1996
8 **Winfried Just and Martin Weese,** Discovering modern set theory. I: The basics, 1996
7 **Gerald J. Janusz,** Algebraic number fields, second edition, 1996
6 **Jens Carsten Jantzen,** Lectures on quantum groups, 1996
5 **Rick Miranda,** Algebraic curves and Riemann surfaces, 1995
4 **Russell A. Gordon,** The integrals of Lebesgue, Denjoy, Perron, and Henstock, 1994
3 **William W. Adams and Philippe Loustaunau,** An introduction to Gröbner bases, 1994
2 **Jack Graver, Brigitte Servatius, and Herman Servatius,** Combinatorial rigidity, 1993
1 **Ethan Akin,** The general topology of dynamical systems, 1993

Classical Groups and Geometric Algebra

GRADUATE STUDIES IN MATHEMATICS 39

Classical Groups and Geometric Algebra

Larry C. Grove

AMERICAN MATHEMATICAL SOCIETY
Providence, Rhode Island

2000 *Mathematics Subject Classification.* Primary 20G15, 20G40, 11E57; Secondary 11E39, 11E88, 51N30.

ABSTRACT. This is a graduate level textbook intended to introduce students to the basic facts about classical groups of linear transformations or matrices from first principles. The main prerequisites are fairly standard courses in linear algebra and abstract algebra.

Library of Congress Cataloging-in-Publication Data

Grove, Larry C.
Classical groups and geometric algebra / Larry C. Grove.
p. cm. — (Graduate studies in mathematics ; v. 39)
Includes bibliographical references and index.
ISBN 0-8218-2019-2 (alk. paper) | Softcover ISBN 1-4704-7974-5
1. Group theory. 2. Geometry, Algebraic. I. Title. II. Series.

QA174.2 .G78 2001
512′.2—dc21 2001046251

Printed in the United States of America.
Reprinted by the American Mathematical Society, 2024.

∞ The paper used in this book is acid-free and falls within the guidelines
established to ensure permanence and durability.
Visit the AMS home page at URL: http://www.ams.org/

11 10 9 8 7 6 5 4 3 2   29 28 27 26 25 24

Contents

Preface

The present volume is intended to be a text for a graduate-level course. It discusses in some detail the groups that are popularly known as the *classical groups*, as they were named by Hermann Weyl [**74**]. They are groups of matrices, or (perhaps more often) quotients of matrix groups by small (typically central) normal subgroups.

The story begins, as Weyl suggested, with *Her All-embracing Majesty*, the *General Linear Group* $\mathrm{GL}(V)$ of all invertible linear transformations of a vector space V over a (commutative) field F. All further groups discussed are either subgroups of $\mathrm{GL}(V)$ or closely related quotient groups.

Most of the classical groups are singled out within Her All-embracing Majesty for basically geometric reasons – they consist of invertible linear transformations that respect a bilinear form having some geometric significance, e.g. a quadratic form (hence preserving "distance"), or a symplectic form, etc. Accordingly, we develop the required geometric notions, albeit from an algebraic point of view, as the end results should apply to vector spaces over more-or-less arbitrary fields, finite or infinite. It is to be hoped that the end result is consonant with the title and intent of Emil Artin's deservedly famous book *Geometric Algebra* [**3**]. In particular, we do not employ Lie-theoretic techniques, important as they are from many other points of view.

The classical groups have proved to be important in a wide variety of venues, ranging from physics to geometry and far beyond. In recent years they have played a prominent role in the classification of the finite simple groups.

There are uniform theories that apply to the classical groups. The theory of Chevalley groups and twisted analogues, based on semisimple Lie algebras, was begun by C. Chevalley [**13**] in 1955, and carried further by R. Steinberg [**65**], J. Tits [**68**], and D. Hertzig [**35**]. Subsequently Tits [**69**] introduced the theory of buildings and groups with BN-pairs, yielding a geometric interpretation of Lie groups and Chevalley groups. For further expositions of these ideas see [**9**], [**11**], [**25**], and [**60**]. The intention in the present volume is rather the opposite, studying the classes of groups one at a time, although we will indicate the identifications with Chevalley groups.

The information contained herein is available from other sources, but it seems not to be easily available in a single easily accessible volume, particularly since [**67**] has gone out of print. Quite simply, we seek to provide a single source for the basic facts about the classical groups defined over fields, together with the required geometrical background information, from first principles. The chief prerequisites are basic linear algebra and abstract algebra, including fundamentals of group theory and some Galois Theory. In fact a fair amount of linear algebra is included in the text, since experience dictates that students' previous exposures to the subject tend to be rather uneven.

Readers familiar with the literature will observe serious debts to a number of excellent sources, most notably works by Artin [**3**], Dieudonné [**21**], Huppert [**41**] and [**42**], and Jacobson [**44**] and [**45**].

Many thanks to Olga Yiparaki for a careful reading of part of the manuscript. Many thanks, as well, to three anonymous referees who offered a number of useful comments and suggestions. The editorial staff of the AMS has been unfailingly helpful; I particularly wish to acknowledge the assistance of Sergei Gelfand, Christine Thivierge, and Arlene O'Sean.

The book was typeset by means of LaTeX 2_ε. The scientific community, and the mathematical community in particular, owes a huge and obvious debt to Donald Knuth [**49**], and subsequently to Leslie Lamport [**51**] and to many others (e.g. [**26**]), who have freed us up to *try* to do things right the first time.

Larry C. Grove

The University of Arizona

June 2001

Chapter 0

Permutation Actions

This chapter is a very brief introduction to permutation group actions, presenting only topics that will re-appear in later chapters. For more detailed discussions of these and other topics see [**10**], [**30**], [**41**], or [**57**], or almost any book on group theory.

It is assumed that the reader is familiar with the elementary group-theoretical material normally covered in a standard first graduate-level algebra course (in the United States). This includes such concepts as normal subgroup, center, derived (commutator) subgroup, simple group, Sylow subgroup, presentation, etc.

The reader is also assumed to be familiar with material from a junior/senior-level linear algebra course, although a fair amount of linear algebra is included in the text as fill-in material, since the content of current linear algebra courses seems not to be very uniform.

Any gaps in either group theory or linear algebra can be filled by browsing in one or more of the many texts available for such courses.

If G is a group and x, $y \in G$ we shall write x^y to denote the conjugate $y^{-1}xy$ and ${}^y x$ to denote yxy^{-1}. Note that $x^{yz} = (x^y)^z$ and ${}^{yz}x = {}^y({}^z x)$ for all x, y, $z \in G$.

If S is a set we write $\mathrm{Perm}(S)$ for the group of all permutations of S, i.e. all bijections from S to S, with composition of functions as the group operation. If G is a group we say that G *acts on* S if there is a homomorphism $\varphi : G \to \mathrm{Perm}(S)$. If φ is one-to-one the action is called *faithful.* Typically G will be a group of linear transformations and S a set of vectors, so if $x \in G$ and $a \in S$ it will be natural to express the action as

$$a \mapsto \varphi(x)a,$$

which is usually abbreviated to

$$a \mapsto xa,$$

the permutation action mapping φ being understood.

With this convention, the definition of a permutation action of G on S reduces to the following two requirements:

$$1a = a \text{ and } (xy)a = x(ya)$$

for all $a \in S$, all $x, y \in G$.

There are various notational conventions in use for group actions – a glance at [**30**], page 3, may be helpful.

Note that the relation on S, whereby a, $b \in S$ are related if $xa = b$ for some $x \in G$, is an equivalence relation. Its equivalence classes are called *orbits*; we write

$$\mathrm{Orb}_G(a) = Ga = \{xa \colon x \in G\}.$$

Thus S is a disjoint union of G-orbits.

For any $a \in S$ define its *stabilizer* to be

$$\mathrm{Stab}_G(a) = \{x \in G \colon xa = a\},$$

a subgroup of G.

The first proposition is fundamental.

Proposition 0.1. *If a group G acts as a permutation group on a set S, and $a \in S$, then*

$$|\,\mathrm{Orb}_G(a)| = [G \colon \mathrm{Stab}_G(a)].$$

Proof. If $H = \mathrm{Stab}_G(a)$ then the mapping $xa \longleftrightarrow xH$ is a bijection between orbit elements and cosets of the stabilizer. □

Say that G acts *transitively* on S if $\mathrm{Orb}_G(a) = S$ for some (and hence for all) $a \in S$. If $|S| \geq 2$ then G acts *doubly transitively* on S if for any pairs (a, b), (c, d) in $S \times S$, with $a \neq b$ and $c \neq d$, there is some $x \in G$ with $xa = c$ and $xb = d$. Note that G is doubly transitive on S if and only if it is transitive and $\mathrm{Stab}_G(a)$ is transitive on $S \setminus \{a\}$ for each $a \in S$.

Suppose that G is transitive on S. A *block (of imprimitivity)* for G in S is a subset $B \subseteq S$, with $|B| \geq 2$, $B \neq S$, such that for each $x \in G$ either $xB = B$ or $xB \cap B = \emptyset$. If G has no blocks in S then G acts *primitively* on S, otherwise it acts *imprimitively*.

Examples

1. Let G be the group of rotational symmetries of a regular tetrahedron in $\mathbb{R}^3$ (it is isomorphic with the alternating group A_4), and let S be the set of four vertices of the tetrahedron. Then G acts primitively on S.
2. Let $G = D_4$, the dihedral group of order 8, acting on the set S of four vertices of a square in $\mathbb{R}^2$. Then G is imprimitive; the set B consisting of two opposite vertices is a block.

Exercises

1. Suppose that G acts transitively on S and $|S| = p$, a prime. Show that G is primitive.
2. Suppose that G is transitive on S. Show that the action is primitive if and only if every $\mathrm{Stab}_G(a)$, $a \in S$, is a maximal subgroup of G.

Proposition 0.2. *If G is doubly transitive on S then G is primitive.*

Proof. Suppose that $B \subseteq S$, $B \neq S$, and $|B| \geq 2$. Choose $a, b \in B$, $a \neq b$, and $c \in S \setminus B$. Choose $x \in G$ with $xa = a$, $xb = c$. Then $a \in xB \cap B$, so $xB \cap B \neq \emptyset$. But $c \in xB \setminus B$, so $xB \neq B$, and B is not a block. □

Proposition 0.3. *Suppose that G acts primitively on S, $N \triangleleft G$, and N is not contained in the kernel of the permutation action of G on S. Then N is transitive on S.*

Proof. Since N is not in the kernel of the action there is an N-orbit in S having two or more elements, say $B = \mathrm{Orb}_N(a)$ and $|B| \geq 2$. If $x \in G$ then $xB = x \cdot Na$, which is equal to $N \cdot xa = \mathrm{Orb}_N(xa)$ by the normality of N in G. Since B and xB are both orbits they are either disjoint or equal. But B cannot be a block for G, so $B = S = \mathrm{Orb}_N(a)$, and N is transitive. □

Proposition 0.4. *Suppose that G acts on S, $N \leq G$, and N is transitive on S. If $a \in S$ then $G = N \cdot \mathrm{Stab}_G(a)$.*

Proof. If $x \in G$ then $xa = ya$ for some $y \in N$. Thus $y^{-1}xa = a$, so $y^{-1}x \in \mathrm{Stab}_G(a)$ and $x \in y \cdot \mathrm{Stab}_G(a) \subseteq N \cdot \mathrm{Stab}_G(a)$. □

A point concerning notation: if S is a subset of a group G we shall, as is customary, write $\langle S \rangle$ to denote the subgroup of G generated by S. It will be convenient to extend this to vector spaces: if X is a subset of a vector space V we shall write $\langle X \rangle$ to denote the sub*space* of V spanned by X. Of course a vector space is also, somewhat incidentally, a group, but the slight ambiguity should never cause any confusion.

The next theorem will be applied many times throughout the book. If G is a group then G' denotes its *derived group*, or *commutator subgroup*. If $G^{(1)}$ is defined to be G', then the *derived series* of G is defined recursively by $G^{(m+1)} = G^{(m)'}$, and G is *solvable* if $G^{(k)} = 1$ for some $k \in \mathbb{N}$.

Theorem 0.5 (Iwasawa 1941 [**43**]). *Suppose that G is faithful and primitive on S and $G' = G$. Fix $s \in S$ and set $H = \mathrm{Stab}_G(s)$. Suppose there is a solvable subgroup $K \triangleleft H$ such that $G = \langle \cup\{K^x \colon x \in G\}\rangle$. Then G is simple.*

Proof. Suppose that $1 \neq N \triangleleft G$. By Proposition 0.3 N is transitive on S, and by Proposition 0.4 $G = HN$. Thus $KN \triangleleft HN = G$. If $x \in G$ then $K^x \leq (KN)^x = KN$, so $\cup\{K^x \colon x \in G\} \subseteq KN$ and hence $KN = G$. Since K is solvable $K^{(m)} = 1$ for some $K^{(m)}$ in the derived series of K. Check inductively that $(KN)^{(\ell)} \leq K^{(\ell)}N$ for all ℓ. Thus

$$G = G^{(m)} = (KN)^{(m)} \leq K^{(m)}N = N$$

and $N = G$. □

Chapter 1

The Basic Linear Groups

We begin with the most basic of the classical groups[1], often called the *linear groups.* Of these, the most basic of the basic is GL(V), the *general linear group*, or *Her All-embracing Majesty*, as she was crowned by Hermann Weyl ([**74**]).

The basic pattern established in this chapter will, with variations, persist throughout the book. We typically begin with a significant subgroup of GL(V), pass to a large subgroup of that subgroup, pass to the quotient of that subgroup by its center, then show that, with few exceptions, the resulting group is simple. If it is also finite we will determine its order, and in some cases compare it with groups from other chapters.

One important reason for carrying out the process described above is that in the end we will have described a significant proportion of all possible finite simple groups. So—let us begin.

If V is an n-dimensional vector space over a field F then GL(V) denotes the *general linear group* of V, i.e. the group of all invertible linear transformations on V. Choosing a basis for V provides an isomorphism of GL(V) with the group GL(n, F) of all invertible $n \times n$ matrices over F.

The determinant function is a homomorphism from GL(V) (or from GL(n, F)) onto the multiplicative group $F^\star = F \setminus \{0\}$; its kernel is the

[1]Chapters 1, 2 and 3 are adapted from sections of *Groups and Characters* [**30**], by L. Grove,

subgroup $\mathrm{SL}(V)$ (or $\mathrm{SL}(n,F)$), the *special linear group*. Thus by definition

$$\mathrm{SL}(V) = \{\tau \in \mathrm{GL}(V)\colon \det \tau = 1\}.$$

If F is finite, with q elements, then the matrix groups are often denoted by $\mathrm{GL}(n,q)$ and $\mathrm{SL}(n,q)$.

Exercise

Show that the centers of $\mathrm{GL}(V)$ and $\mathrm{SL}(V)$ are

$$Z(\mathrm{GL}(V)) = \{a1\colon a \in F^\star\} \text{ and } Z(\mathrm{SL}(V)) = \{a1\colon a \in F^\star,\ a^n = 1\}.$$

If $0 \neq v \in V$ write $[v]$ for the line $Fv = \langle v \rangle$ through the origin spanned by v, and call it a *projective point.* The set of all distinct projective points $[v]$ is called the *projective space* of dimension $n-1$ based on V, and is denoted by $\mathbb{P}_{n-1}(V)$, or simply by $\mathbb{P}(V)$. There is a natural permutation action of $\mathrm{GL}(V)$ on $\mathbb{P}(V)$, given by $\tau[v] = [\tau v]$ for all $\tau \in \mathrm{GL}(V)$, $[v] \in \mathbb{P}(V)$. It is easy to check that the kernel of the action is $Z(\mathrm{GL}(V))$, and likewise that the kernel of the action of $\mathrm{SL}(V)$ on $\mathbb{P}(V)$ is $Z(\mathrm{SL}(V))$.

Exercise

If $\dim V = n$ and $|F| = q$, finite, show that $|\mathbb{P}(V)| = (q^n - 1)/(q-1)$.

Define the *projective general linear group* of V to be

$$\mathrm{PGL}(V) = \mathrm{GL}(V)/Z(\mathrm{GL}(V))$$

and the *projective special linear group* to be

$$\mathrm{PSL}(V) = \mathrm{SL}(V)/Z(\mathrm{SL}(V));$$

each of them acts faithfully on $\mathbb{P}(V)$. There are isomorphic copies based on the corresponding matrix groups, with obvious corresponding notation.

Proposition 1.1. *If $|F| = q$, finite, then*

1. $|\,\mathrm{GL}(n,q)| = q^{n(n-1)/2}\prod\{q^k - 1\colon 1 \le k \le n\}$,
2. $|\,\mathrm{SL}(n,q)| = |\,\mathrm{PGL}(n,q)| = |\,\mathrm{GL}(n,q)|/(q-1)$, *and*
3. $|\,\mathrm{PSL}(n,q)| = |\,\mathrm{SL}(n,q)|/(n, q-1)$, *where $(n, q-1)$ denotes, as usual, the greatest common divisor of n and $q-1$.*

Proof. The first row of a matrix in $\mathrm{GL}(n,q)$ can be any of the $q^n - 1$ nonzero row vectors over F; the second must be independent of the first, so there are $q^n - q$ choices, etc. Thus

$$\begin{aligned} |\,\mathrm{GL}(n,q)| &= \prod_1^n (q^n - q^k) = q^{1+\cdots+(n-1)}\prod_1^n (q^k - 1) \\ &= q^{n(n-1)/2}\prod\{q^k - 1\colon 1 \le k \le n\}. \end{aligned}$$

Part 2 follows, since $|F^\star| = q-1$, $Z(\mathrm{GL}(n,q)) \cong F^\star$, and

$$\mathrm{GL}(n,q)/\mathrm{SL}(n,q) \cong F^\star.$$

Since $F^\star$ is a cyclic group it follows from the exercise on page 6 that $a \cdot 1 \in Z(\mathrm{SL}(n,q))$ if and only if $a^n = a^{q-1} = 1$, which is if and only if $a^{(n,q-1)} = 1$. In other words, a must be in the unique subgroup of order $(n, q-1)$ in $F^\star$. □

Although SL and PGL have the same order they are usually *not* isomorphic.

Exercise

If F is any field denote by $\mathcal{M}_2(F)$ the group of all *linear fractional transformations* (sometimes called *Möbius transformations*) of F, i.e. all rational functions $f(x) \in F(x)$ of the form $f(x) = \dfrac{ax+b}{cx+d}$, with $ad-bc \neq 0$. Show that $\mathcal{M}_2(F) \cong \mathrm{PGL}(2,F)$.

Let $\mathcal{L}_2(F)$ be the subgroup of $\mathcal{M}_2(F)$ consisting of those $f(x)$ for which $ad-bc$ is a square in $F^\star$. Show that $\mathcal{L}_2(F) \cong \mathrm{PSL}(2,F)$.

A *hyperplane* in V is any subspace of codimension 1. If $1 \neq \tau \in \mathrm{GL}(V)$ then τ is called a *transvection* if there is a hyperplane W such that $\tau|_W = 1_W$ and $\tau v - v \in W$ for all $v \in V$; W is called the *fixed hyperplane* of τ.

If τ is a transvection with fixed hyperplane W choose a basis for V consisting of first some $v_1 \in V \setminus W$ and then a basis $\{v_2, \ldots, v_n\}$ for W. It is clear from the matrix representing τ relative to this basis that $\det \tau = 1$, so in fact $\tau \in \mathrm{SL}(V)$.

Exercises

1. Show that the inverse of a transvection is a transvection.
2. Suppose that V is a subspace of a space V_1, $v \in V_1 \setminus V$, and τ is a transvection on V with fixed hyperplane W. Show that τ can be extended to a transvection τ_1 on V_1 whose fixed hyperplane W_1 contains v.

Proposition 1.2. *If u and v are linearly independent in V then there is a transvection τ with $\tau u = v$.*

Proof. Choose a hyperplane W in V with $u - v \in W$ but $u \notin W$, and define τ by means of $\tau|_W = 1_W$, $\tau u = v$. If $x \in V$ write $x = au + w$, where $a \in F$ and $w \in W$. Then $\tau x - x = av + w - au - w = a(v-u) \in W$, so τ is a transvection. □

Proposition 1.3. *Suppose that W_1 and W_2 are two distinct hyperplanes in V and that $v \in V \setminus (W_1 \cup W_2)$. Then there is a transvection τ with $\tau W_1 = W_2$ and $\tau v = v$.*

Proof. Note that $W_1 + W_2 = V$, so $\dim W_1 \cap W_2 = n - 2$ and $W = W_1 \cap W_2 + Fv$ is another hyperplane. Write $v = x + y$, with $x \in W_1$ and $y \in W_2$. Then $x \notin W_2$, so $W_1 = W_1 \cap W_2 + Fx$, and likewise $W_2 = W_1 \cap W_2 + Fy$. Thus $V = W_1 \cap W_2 + Fx + Fy$. It follows that $x \notin W$, or else $y = v - x$ is also in W and hence $V \subseteq W$!! Define τ via $\tau|_W = 1_W$ and $\tau x = y$. Then τ is a transvection just as in the proof above, $\tau v = v$ since $v \in W$, and $\tau W_1 = \tau(W_1 \cap W_2 + Fx) = W_1 \cap W_2 + Fy = W_2$. □

Theorem 1.4. *The set of transvections generates* $\mathrm{SL}(V)$.

Proof. Fix $\rho \in \mathrm{SL}(V)$, then choose a hyperplane W in V and choose a vector $v \in V \setminus W$. If v and ρv are linearly independent then by Proposition 1.2 there is a transvection τ_1 with $\tau_1 \rho v = v$. If v and ρv are linearly dependent then first choose a transvection τ_0 so that v and $\tau_0 \rho v$ are linearly independent, then a transvection τ_1' so that $\tau_1' \tau_0 \rho v = v$, and set $\tau_1 = \tau_1' \tau_0$. Thus in either case we have $\tau_1 \rho v = v$ and τ_1 is a product of transvections. Note that $v \notin \tau_1 \rho W$. If $\tau_1 \rho W = W$ set $\tau_2 = 1_V$. If $\tau_1 \rho W \neq W$ apply Proposition 1.3 to get a transvection τ_2 with $\tau_2 \tau_1 \rho W = W$ and $\tau_2 v = v$. Set $\sigma = \tau_2 \tau_1 \rho$. Since $\sigma v = v$ it follows that $\sigma|_W \in \mathrm{SL}(W)$. Now use induction on $n = \dim V$. If $n = 2$ then $\sigma|_W = 1_W$, so $\sigma = 1$ and $\rho = \tau_1^{-1} \tau_2^{-1}$. If $n > 2$ then by induction $\sigma|_W$ is a product of transvections on W, each of which extends to a transvection fixing v on V by Exercise 2 on page 7. Thus σ is the product of the extended transvections, and $\rho = \tau_1^{-1} \tau_2^{-1} \sigma$, a product of transvections. □

Proposition 1.5. *If τ_1 and τ_2 are transvections on V then they are conjugate in* $\mathrm{GL}(V)$. *If $n > 2$ they are conjugate in* $\mathrm{SL}(V)$.

Proof. For $i = 1, 2$ write W_i for the fixed hyperplane of τ_i, choose $x_i \in V \setminus W_i$, and set $w_i = \tau_i x_i - x_i \in W_i$. Choose bases $\{w_1, u_3, \dots u_n\}$ for W_1 and $\{w_2, v_3, \dots, v_n\}$ for W_2. For each $a \in F^\star$ define $\sigma_a \in \mathrm{GL}(V)$ by setting $\sigma_a x_1 = x_2$, $\sigma_a w_1 = w_2$, $\sigma_a u_i = v_i$ for $3 \le i \le n-1$, and, if $n > 2$, $\sigma_a u_n = a v_n$. Then a straightforward calculation verifies that $\sigma_a \tau_1 \sigma_a^{-1}$ and τ_2 agree on $x_2, w_2, u_3, \dots, u_n$, and hence $\sigma_a \tau_1 \sigma_a^{-1} = \tau_2$.

If $n > 2$ we may set $b = (\det \sigma_1)^{-1}$ and obtain $\sigma_b \in \mathrm{SL}(V)$. □

Proposition 1.6. *Suppose that* $\dim V = 2$, *and let $\{v_1, v_2\}$ be any basis for V. Every transvection is conjugate in* $\mathrm{SL}(V)$ *to one whose matrix relative to $\{v_1, v_2\}$ is of the form* $\begin{bmatrix} 1 & 0 \\ a & 1 \end{bmatrix}$, $a \in F^\star$.

Proof. If τ is a transvection with fixed hyperplane (line) W choose $v \in V \setminus W$ and set $w = \tau v - v \in W$. Relative to the basis $\{v, w\}$ the matrix representing τ is $\begin{bmatrix} 1 & 0 \\ 1 & 1 \end{bmatrix}$. If M is the matrix that represents τ relative to $\{v_1, v_2\}$ then there is a matrix B in $\mathrm{GL}(2, F)$ with $B^{-1}MB = \begin{bmatrix} 1 & 0 \\ 1 & 1 \end{bmatrix}$. If $\det B = a^{-1}$ we may set $A = \begin{bmatrix} a & 0 \\ 0 & 1 \end{bmatrix}$. Then $BA \in \mathrm{SL}(2, F)$, and

$$(BA)^{-1}M(BA) = A^{-1}\begin{bmatrix} 1 & 0 \\ 1 & 1 \end{bmatrix}A = \begin{bmatrix} 1 & 0 \\ a & 1 \end{bmatrix}.$$

□

Theorem 1.7. *If $n \geq 3$ and $G = \mathrm{SL}(V)$ then $G' = G$.*

Proof. By Theorem 1.4 and Proposition 1.5 it will suffice to exhibit a transvection in G'. Choose a basis $\{v_1, \ldots, v_n\}$ for V and define τ_1, τ_2 via

$$\tau_1 \colon v_1 \mapsto v_1 - v_2,\ v_i \mapsto v_i \ \text{ if } 2 \leq i \leq n,$$

$$\tau_2 \colon v_1 \mapsto v_1,\ v_2 \mapsto v_2 - v_3,\ v_i \mapsto v_i \ \text{ if } 3 \leq i \leq n.$$

Then

$$\tau_1\tau_2\tau_1^{-1}\tau_2^{-1} \colon v_1 \mapsto v_1 - v_3,\ v_i \mapsto v_i \ \text{ if } 2 \leq i \leq n,$$

so $\tau_1\tau_2\tau_1^{-1}\tau_2^{-1}$ is a transvection in G'. □

Corollary 1.8. *If $n \geq 3$ then $\mathrm{PSL}(V)$ is equal to its derived group.*

Theorem 1.9. *If $n = 2$, $|F| > 3$, and $G = \mathrm{SL}(V)$, then $G' = G$.*

Proof. Choose a basis $\{v_1, v_2\}$ for V and choose $a \in F^\star$, $a \neq \pm 1$. Define $\sigma \in \mathrm{SL}(V)$ via $\sigma(v_1) = a^{-1}v_1$, $\sigma(v_2) = av_2$, and for each $b \in F^\star$ define $\tau_b \in \mathrm{SL}(V)$ via $\tau_b(v_1) = v_1 + bv_2$, $\tau_b(v_2) = v_2$. Then $\sigma\tau_b\sigma^{-1}\tau_b^{-1}$ is represented by the matrix

$$\begin{bmatrix} a^{-1} & 0 \\ 0 & a \end{bmatrix}\begin{bmatrix} 1 & 0 \\ b & 1 \end{bmatrix}\begin{bmatrix} a & 0 \\ 0 & a^{-1} \end{bmatrix}\begin{bmatrix} 1 & 0 \\ -b & 1 \end{bmatrix} = \begin{bmatrix} 1 & 0 \\ ba(a - a^{-1}) & 1 \end{bmatrix}.$$

The theorem follows from Theorem 1.4 and Proposition 1.6, since $b \in F^\star$ is arbitrary. □

Corollary 1.10. *If $n = 2$ and $|F| > 3$ then $\mathrm{PSL}(V)$ is equal to its derived group.*

Exercise

If $n \geq 3$ or $|F| > 3$ show that the derived group of $\mathrm{GL}(V)$ is $\mathrm{SL}(V)$. Determine the derived groups of $\mathrm{SL}(2,2)$, $\mathrm{SL}(2,3)$, and $\mathrm{GL}(2,3)$.

Proposition 1.11. *If $n \geq 2$ then $\mathrm{PSL}(V)$ acts faithfully and doubly transitively on the projective space $\mathbb{P}(V)$. In particular it is primitive.*

Proof. It was observed earlier that $\mathrm{PSL}(V)$ is faithful on $\mathbb{P}(V)$. Take $[v_1] \neq [v_2]$ and $[w_1] \neq [w_2]$ in $\mathbb{P}(V)$, so $\{v_1, v_2\}$ and $\{w_1, w_2\}$ are linearly independent sets in V. If $n = 2$ they are bases. If $n \geq 3$ set $V_1 = \langle v_1, v_2 \rangle$ and $V_2 = \langle w_1, w_2 \rangle$. Then either $V_1 = V_2 \neq V$ or else $V_1 \neq V_2$, in which case $V_1 \cup V_2$ is not a subspace – in either case $V_1 \cup V_2 \neq V$. If $v_3 \in V \setminus (V_1 \cup V_2)$ then both $\{v_1, v_2, v_3\}$ and $\{w_1, w_2, v_3\}$ are linearly independent. The argument may be repeated to obtain two bases for V, $\{v_1, v_2, v_3, \ldots, v_n\}$ and $\{w_1, w_2, v_3, \ldots, v_n\}$. For any $b \in F^\star$ define $\tau_b \in \mathrm{GL}(V)$ via $v_1 \mapsto bw_1$, $v_2 \mapsto w_2$, $v_i \mapsto v_i$ for $3 \leq i \leq n$. If $w_j = \sum_i a_{ij} v_i$, $j = 1, 2$, then $\det \tau_b = b(a_{11}a_{22} - a_{12}a_{21})$. Choose b so that $\det \tau_b = 1$; then $\tau_b \in \mathrm{SL}(V)$, and it carries $[v_1]$ to $[w_1]$, $[v_2]$ to $[w_2]$. □

Proposition 1.12. *If $0 \neq v \in V$ and $A = \mathrm{Stab}_{SL(V)}([v])$ then A has an abelian normal subgroup B whose conjugates in $\mathrm{SL}(V)$ generate $\mathrm{SL}(V)$.*

Proof. Choose a hyperplane W with $v \notin W$, so $V = W + Fv$. If $\sigma \in A$ and $w \in W$ write $\sigma w = \sigma' w + a_w v$, with $\sigma' w \in W$ and $a_w \in F$. It is easy to check that $\sigma' \in \mathrm{GL}(W)$ and that $\varphi \colon \sigma \mapsto \sigma'$ is a homomorphism from A into $\mathrm{GL}(W)$. Thus $B = \mathrm{Ker}\ \varphi$ is a normal subgroup of A.

Choose a basis $\{w_1, \ldots, w_{n-1}\}$ for W. For any $b \in F^\star$ define $\tau_b \in A$ via $w_1 \mapsto w_1 + bv$, $w_i \mapsto w_i$ for $2 \leq i \leq n-1$, and $v \mapsto v$. Then τ_b is a transvection (with hyperplane spanned by $\{w_2, \ldots, w_{n-1}, v\}$), and if $w \in W$ then $\tau_b w = w + bv$, so $\tau_b' = 1_W$ and $\tau_b \in B$. If $n = 2$ the matrix representing τ_b relative to the basis $\{w_1, v\}$ is $\begin{bmatrix} 1 & 0 \\ b & 1 \end{bmatrix}$, and $b \in F^\star$ is arbitrary. It follows from Theorem 1.4 and Propositions 1.5 and 1.6 that the conjugates of B generate $\mathrm{SL}(V)$ for all $n \geq 2$. If $\sigma \in B$ then $\sigma' = 1_W$, so $\sigma w_i = w_i + a_i v$, all i, $a_i \in F$. Thus, if $\sigma_1, \sigma_2 \in B$, their representing matrices relative to the basis $\{w_1, \ldots, w_{n-1}, v\}$ have the partitioned form

$$\begin{bmatrix} I & 0 \\ u_1 & 1 \end{bmatrix} \quad \text{and} \quad \begin{bmatrix} I & 0 \\ u_2 & 1 \end{bmatrix}.$$

Since

$$\begin{bmatrix} I & 0 \\ u_1 & 1 \end{bmatrix}\begin{bmatrix} I & 0 \\ u_2 & 1 \end{bmatrix} = \begin{bmatrix} I & 0 \\ u_1 + u_2 & 1 \end{bmatrix} = \begin{bmatrix} I & 0 \\ u_2 & 1 \end{bmatrix}\begin{bmatrix} I & 0 \\ u_1 & 1 \end{bmatrix}$$

we see that B is abelian. □

Theorem 1.13. *If* $\dim V = n \geq 2$, *then every* $\mathrm{PSL}(V)$ *is a simple group, except for* $\mathrm{PSL}(2,2)$ *and* $\mathrm{PSL}(2,3)$.

Proof. Choose $v \neq 0$ in V, and take $A = \mathrm{Stab}_{\mathrm{SL}(V)}([v])$ and $B \lhd A$ as in Proposition 1.12. Write $Z = Z(\mathrm{SL}(V))$ and set $H = A/Z = \mathrm{Stab}_{\mathrm{PSL}(V)}([v])$, $K = BZ/Z \lhd H$. Then all the conditions of Iwasawa's Theorem (0.5) are met, and $\mathrm{PSL}(V)$ is simple. □

$\mathrm{PSL}(n, F)$ is the Chevalley group $A_{n-1}(F)$ (see Carter [**11**], page 184).

Exercise

Use the action on the projective line $\mathbb{P}$ to show that $\mathrm{PSL}(2,2) \cong \mathrm{Sym}(3)$, $\mathrm{PSL}(2,3) \cong \mathrm{Alt}(4)$, and $\mathrm{PSL}(2,4) \cong \mathrm{Alt}(5)$. Note also that $\mathrm{PSL}(2,5) \cong \mathrm{Alt}(5)$ (why?).

Note that $|\mathrm{PSL}(3,4)| = 20160 = |\mathrm{Alt}(8)|$, and that both groups are simple. Let us show that they are *not* isomorphic. For any $\sigma \in \mathrm{SL}(3,4)$ write $\overline{\sigma}$ for its image in PSL.

If $\overline{\sigma}$ is any element of order 2 in $\mathrm{PSL}(3,4)$ then $\sigma^2 \in Z(\mathrm{SL}(3,4))$, so $\sigma^6 = 1$. Replacing σ by $\tau = \sigma^3$ we have $\overline{\tau} = \overline{\sigma}$ and τ of order 2 in SL as well as in PSL. Let $W = \mathrm{Ker}(1+\tau) \subseteq V$, so $\dim V = 3 = \dim W + \dim (1+\tau)V$. Note that $v \in W$ if and only if $\tau v = v$ (characteristic 2), so $W \neq V$. Also $\tau(1+\tau)v = (\tau+1)v$, all $v \in V$; i.e. $(1+\tau)v$ is fixed by τ. Thus $(1+\tau)V \subseteq W$. Now $\dim (1+\tau)V \leq \dim W$, and it follows that $\dim (1+\tau)V = 1$, $\dim W = 2$. Consequently W is a hyperplane and $\tau v - v \in W$ for all $v \in V$; i.e. τ is a transvection. By Proposition 1.5 we see that all elements of order 2 are conjugate in $\mathrm{PSL}(3,4)$.

Since $\mathrm{Alt}(8)$ has nonconjugate elements of order 2, e.g. $(1\,2)(3\,4)$ and $(1\,2)(3\,4)(5\,6)(7\,8)$, it is immediate that $\mathrm{PSL}(3,4) \not\cong \mathrm{Alt}(8)$.

Exercise

Note also that $\mathrm{PSL}(4,2)$ and $\mathrm{PSL}(3,4)$ are of the same order. Show that $\mathrm{PSL}(4,2)$ and $\mathrm{PSL}(3,4)$ are not isomorphic by considering the matrices

$$\begin{bmatrix} 1 & 0 & 0 & 0 \\ 0 & 1 & 0 & 0 \\ 0 & 0 & 1 & 0 \\ 0 & 0 & 1 & 1 \end{bmatrix} \quad \text{and} \quad \begin{bmatrix} 1 & 0 & 0 & 0 \\ 1 & 1 & 0 & 0 \\ 0 & 0 & 1 & 0 \\ 0 & 0 & 1 & 1 \end{bmatrix}$$

in $\mathrm{SL}(4,2)$.

It is in fact true that $\mathrm{PSL}(4,2) \cong \mathrm{Alt}(8)$. One proof proceeds by finding a set of generators for $\mathrm{PSL}(4,2)$ that satisfy the relations in a presentation for $\mathrm{Alt}(8)$; see Huppert [**41**], page 157, for details.

It can be shown that any simple group of order 168 is isomorphic to PSL(2, 7) (see [**41**], page 184). That applies in particular to PSL(3, 2), the collineation group of the world's smallest projective plane P, with 7 points and 7 lines.

In fact the simple group of order 168 has a remarkable history. It is the symmetry group of a quartic surface studied by Felix Klein, the locus $\mathcal{X}$ in $\mathbb{P}^2$ of the homogeneous equation $X^3Y + Y^3Z + Z^3X = 0$. Its study has ramifications ranging from Fermat's Last Theorem to imaginary quadratic fields to classical modular curves. Readers are referred to the fascinating account of the history by N. Elkies ([**24**]).

Finally, $|\,\mathrm{PSL}(2,9)| = 360 = |\,\mathrm{Alt}(6)|$. In this case PSL(2, 9) has a subgroup isomorphic with PSL(2, 5), hence of index 6, so the action on its cosets gives an isomorphism into Sym(6), and $\mathrm{PSL}(2,9) \cong \mathrm{Alt}(6)$ since it is the unique subgroup of index 2 of Sym(6). Again, see [**41**] for details.

Chapter 2

Bilinear Forms

All further groups to be discussed will be subgroups (or quotients thereof) of $\mathrm{GL}(V)$ determined by geometric considerations. The relevant geometry is generally determined by some sort of bilinear or sesquilinear form on V.

It should perhaps be noted that the linear groups of the previous chapter can fit into the same context, albeit in a somewhat artificial fashion, by considering them as symmetry groups of the trivial bilinear form $B(x, y) = 0$ for all $x, y \in V$.

We begin the study of (usually nontrivial) bilinear forms in this chapter, and carry the investigations further as needed in later chapters.

Suppose that F is a field and V is a vector space over F. Then its *dual space*, $V^* = \mathrm{Hom}_F(V)$, is the set of all *linear functionals* from V to F; it is also an F-vector space. If $\{v_1, \ldots, v_n\}$ is a basis for V define $\{v_1^*, \ldots, v_n^*\} \subseteq V^*$ via $v_i^*(v_j) = \delta_{ij}$, all i and j (and of course extend by linearity). This defines the *dual basis* for V^*. If $\sum_i a_i v_i^* = 0$ then $0 = \sum_i a_i v_i^*(v_j) = a_j$, all j; if $f \in V^*$ set $f(v_i) = b_i$ and check that $f = \sum_i b_i v_i^*$. In particular $\dim_F(V^*) = \dim_F(V)$ (when finite).

A *bilinear form* on V is a function $B\colon V \times V \to F$ that becomes a linear functional in either variable if the other variable is fixed, i.e.

$$\begin{aligned} B(u+v, w) &= B(u, w) + B(v, w) \text{ and } B(au, w) = aB(u, w); \\ B(u, v+w) &= B(u, v) + B(u, w) \text{ and } B(u, aw) = aB(u, w) \end{aligned}$$

for all $u, v, w \in V$, all $a \in F$.

If B is a bilinear form on V and $\{v_1, \ldots, v_n\}$ is a basis for V set $b_{ij} = B(v_i, v_j)$, all i, j. Then $\widehat{B} = [b_{ij}]$ is called the matrix of B relative to $\{v_i\}$.

If it becomes necessary to exhibit the dependence on the basis we will write $\widehat{B}_{\{v_i\}}$. If $u, w \in V$ write $u = \sum_i a_i v_i$ and $w = \sum_i c_i v_i$, so u and w are represented by column vectors (coordinate vectors) $\widehat{u} = (a_1, \dots, a_n)^t$ and $\widehat{w} = (c_1, \dots, c_n)^t$. Then

$$B(v, w) = B(\sum_i a_i v_i, \sum_j c_j v_j) = \sum_{i,j} a_i B(v_i, v_j) c_j = \widehat{v}^{\,t} \widehat{B} \widehat{w}.$$

Conversely, if $\widehat{B} = [b_{ij}]$ is any $n \times n$ matrix over F we may define a bilinear form (depending on a choice $\{v_i, \dots, v_n\}$ of basis) by decreeing that $B(v_i, v_j) = b_{ij}$ and extending by linearity.

If $\{w_1, \dots, w_n\}$ is another basis write $w_j = \sum_i d_{ij} v_i$, $d_{ij} \in F$, for all j. Then $B(w_i, w_j) = \sum_{k,\ell} d_{ki} B(v_k, v_\ell) d_{\ell,j} = \sum_{k,\ell} d_{ki} b_{k\ell} d_{\ell,j}$, the ij-entry of $D^t \widehat{B} D$, where $D = [d_{ij}]$ is the (invertible) change-of-basis matrix. In particular, any two representing matrices have the same rank, which we define to be the *rank* of the form B.

Classically, square matrices M and N are called *congruent* if $M = D^t N D$ for some invertible matrix D, so two representing matrices for a bilinear form B are congruent. Note that then $\det M = (\det D)^2 \det N$.

Write $F^{\times 2} = \{a^2 \colon a \in F^\star\}$, the subgroup of squares in the multiplicative group $F^\star$. If B is a bilinear form on V and $\widehat{B}$ is a representing matrix define the *discriminant* of B to be

$$\operatorname{discr}(B) = \begin{cases} 0 & \text{if } \det \widehat{B} = 0, \\ (\det \widehat{B}) F^{\times 2} \in F^\star / F^{\times 2} & \text{otherwise.} \end{cases}$$

Note that $\operatorname{discr}(B)$ is independent of the choice of basis. Say that B is *nondegenerate* if $\operatorname{discr}(B) \neq 0$.

If B is a bilinear form on V define two maps L and R from V to V^* as follows: $L\colon v \mapsto L_v$, $R\colon v \mapsto R_v$, with $L_v(w) = B(v, w)$ and $R_v(w) = B(w, v)$, all $w \in V$. It is easy to check that L and R are linear transformations.

Define the *left* and *right radicals* of V (relative to B) to be

$$\operatorname{rad}_L(V) = \operatorname{Ker} L = \{v \in V \colon B(v, w) = 0, \text{ all } w \in V\},$$

$$\operatorname{rad}_R(V) = \operatorname{Ker} R = \{v \in V \colon B(w, v) = 0, \text{ all } w \in V\}.$$

Proposition 2.1. *A bilinear form B on V is nondegenerate if and only if $\operatorname{rad}_L(V) = 0$ if and only if $\operatorname{rad}_R(V) = 0$.*

Proof. Choose a basis $\{v_1, \dots, v_n\}$ and write $\widehat{B} = [b_{ij}]$ for the matrix representing B. Note that $v \in \operatorname{rad}_R(V)$ if and only if $B(v_i, v) = 0$, all i. Write $v = \sum_j a_j v_j$, $a_j \in F$. Then $B(v_i, v) = \sum_j b_{ij} a_j = 0$, all i, if and only if the column vector $(a_1, \dots, a_n)^t$ is a solution to the homogeneous system

$\widehat{B}X = 0$. Thus $\mathrm{rad}_R(V) = 0$ if and only if det $\widehat{B} \neq 0$, and the equivalence with $\mathrm{rad}_L(V) = 0$ is similar. □

Corollary 2.2. *A bilinear form B on V is nondegenerate if and only if $V^* = \mathrm{Im}\ L = \mathrm{Im}\ R$, i.e. if and only if given any $f \in V^*$ there exist u, $v \in V$ such that $f(w) = B(w, u) = B(v, w)$, all $w \in V$.*

The corollary generalizes.

Corollary 2.3. *Suppose that B is a nondegenerate bilinear form on V and that W is a subspace of V. If $f \in W^*$ then there exist u, $v \in V$ such that $f = R_u|_W = L_v|_W$.*

Proof. We may choose a basis $\{w_1, \ldots, w_k\}$ for W and enlarge it to a basis $\{w_1, \ldots, w_n\}$ for V. Extend f to $f_1 \in V^*$ via $f_1|_W = f$, $f_1(v_i) = 0$ for $i > k$. By Corollary 2.2 there exist $u, v \in V$ with $f_1 = R_u = L_v$, and hence $f = R_u|_W = L_v|_W$. □

For any subset $S \subseteq V$ define

$$\perp_L (S) = \{v \in V \colon B(v, w) = 0, \text{ all } w \in S\},$$

$$\perp_R (S) = \{v \in V \colon B(w, v) = 0, \text{ all } w \in S\}.$$

Thus $v \in \perp_L (S)$ if and only if $L_v|_S = 0$, $v \in \perp_R (S)$ if and only if $R_v|_S = 0$. Note that $\perp_L (V) = \mathrm{rad}_L(V)$ and $\perp_R (V) = \mathrm{rad}_R(V)$.

Exercise

Show that

1. $\perp_L (S)$ and $\perp_R (S)$ are subspaces of V,
2. $\perp_L (\perp_R (S)) \supseteq S$ and $\perp_R (\perp_L (S)) \supseteq S$, and
3. if $S \subseteq T$ then $\perp_L (T) \subseteq \perp_L (S)$ and $\perp_R (T) \subseteq \perp_R (S)$.

Proposition 2.4. *If B is a nondegenerate bilinear form on V and W is a subspace of V then*

$$\dim \perp_L (W) = \dim \perp_R (W) = \dim V - \dim W.$$

Proof. The map $\theta \colon v \mapsto L_v|_W$ is a linear transformation from V onto W^* (Corollary 2.3), and Ker $\theta = \perp_L (W)$ by definition. Thus $\dim V = \dim W^* + \dim \perp_L (W)$, and similarly $\dim V = \dim W^* + \dim \perp_R (W)$. □

Corollary 2.5. *If B is a nondegenerate bilinear form on V and W is a subspace of V then $\perp_L (\perp_R (W)) = \perp_R (\perp_L (W)) = W$.*

Proof. By the exercise above $\perp_L(\perp_R(W)) \supseteq W$, and

$$\begin{aligned}\dim \perp_L(\perp_R(W)) &= \dim V - \dim \perp_R(W)\\ &= \dim V - (\dim V - \dim W) = \dim W.\end{aligned}$$

□

Exercise

If B is nondegenerate on V and S is any subset of V show that

$$\perp_L(\perp_R(S)) = \perp_R(\perp_L(S))$$

is the subspace spanned by S.

Suppose that B is a bilinear form on V. Say that B is *symmetric* if $B(v,w) = B(w,v)$ for all $v, w \in V$. Say that B is *alternate* if $B(v,v) = 0$ for all $v \in V$. Note that B is symmetric if and only if any representing matrix $\widehat{B}$ is symmetric, i.e. $\widehat{B}^t = \widehat{B}$.

If B is an alternate form and $v, w \in V$ then $0 = B(v+w, v+w) = B(v,v)+B(v,w)+B(w,v)+B(w,w)$, so $B(w,v) = -B(v,w)$. If char $F = 2$ this shows that B is symmetric; in general it shows that B is *skew symmetric* (also called *antisymmetric*). If char $F \neq 2$ then the concepts of alternate form and skew symmetric form are equivalent, for then skew symmetry entails $B(v,v) = -B(v,v)$, or $B(v,v) = 0$. In general, a representing matrix $\widehat{B} = [b_{ij}]$ for an alternate form is an *alternate*, or *skew symmetric*, matrix, meaning that $b_{ii} = 0$, all i, and $\widehat{B}^t = -\widehat{B}$.

Proposition 2.6. *Suppose that B is a bilinear form on V satisfying*

$$B(u,v)B(w,u) = B(v,u)B(u,w) \qquad (*)$$

for all $u, v, w \in V$. Then B is symmetric or alternate.

Proof. Take $u = v$ and conclude that

$$B(v,v)[B(w,v) - B(v,w)] = 0 \qquad (**)$$

for all $v, w \in V$. We wish to show that either $B(v,v) = 0$ for *all* $v \in V$, or $B(w,v) = B(v,w)$ for *all* $v, w \in V$. If that is not the case then there exist $x, y, z \in V$ with $B(y,y) \neq 0$ and $B(x,z) \neq B(z,x)$. Apply $(**)$ and obtain (i) $B(x,x) = B(z,z) = 0$, (ii) $B(x,y) = B(y,x)$, and (iii) $B(y,z) = B(z,y)$. Use $(*)$, with $u = x$, $v = y$, and $w = z$, to conclude from (ii) that $B(x,y) = B(y,x) = 0$. Interchange u and w in $(*)$ and get, again with $u = x$, $v = y$, and $w = z$, that $B(z,y) = B(y,z) = 0$ from (iii). Thus $B(x, y+z) = B(x,z) \neq B(z,x) = B(y+z, x)$, and, by $(**)$,

$$\begin{aligned}0 &= B(y+z, y+z)\\ &= B(y,y) + B(y,z) + B(z,y) + B(z,w)\\ &= B(y,y) \neq 0,\end{aligned}$$

a contradiction. □

For any bilinear form B on V, and $v, w \in V$, say that v is *orthogonal* to w, and write $v \perp w$, if $B(v, w) = 0$. If orthogonality is a reflexive relation on V (i.e. $v \perp w \Rightarrow w \perp v$) then the form B is called *reflexive*. Note that if B is a reflexive form then $\perp_L (S) = \perp_R (S)$ for all subsets $S \subseteq V$; we agree to write $S^\perp = \perp_L (S) = \perp_R (S)$ in that case. If W is a subspace then $W^\perp$ is called the *orthogonal complement* of W. Note that $W \cap W^\perp = 0$ if and only if $B|_{W \times W}$ is a nondegenerate form on W. If $W \cap W^\perp = 0$ we say that W is a *nondegenerate subspace* (relative to B).

In general, if B is a reflexive bilinear form on V and W is a subspace we define the *radical* of W to be $\text{rad } W = W \cap W^\perp$.

Exercise

Suppose that B is a reflexive bilinear form on V, and $V = U \oplus W$, with $U \perp W$. Show that $\text{rad } V = \text{rad } U \oplus \text{rad } W$.

Proposition 2.7. *A bilinear form B on V is reflexive if and only if it is either symmetric or alternate.*

Proof. $\Leftarrow$: A symmetric form is clearly reflexive. If B is alternate it is skew symmetric, so $v \perp w$ means $0 = B(v, w) = -B(w, v)$, and therefore $w \perp v$.

$\Rightarrow$: Take any $u, v, w \in V$ and set $x = B(u, v)w - B(u, w)v$. Then $B(u, x) = B(u, v)B(u, w) - B(u, w)B(u, v) = 0$, so $u \perp x$. Thus also $x \perp u$, which says $0 = B(x, u) = B(u, v)B(w, u) - B(v, u)B(u, w)$. Apply Proposition 2.6. □

Suppose that B_1, B_2 are bilinear forms on spaces V_1, V_2, respectively. An *isometry* relative to B_1 and B_2 is an F-isomorphism $\sigma: V_1 \to V_2$ satisfying $B_2(\sigma v, \sigma w) = B_1(v, w)$ for all $v, w \in V_1$. If an isometry exists the forms are called *equivalent*.

Proposition 2.8. *Bilinear forms B_1, B_2 on spaces V_1, V_2 are equivalent if and only if there are bases for V_1, V_2 relative to which $\widehat{B}_1 = \widehat{B}_2$.*

Proof. $\Rightarrow$: Let $\sigma: V_1 \to V_2$ be an isometry. Choose a basis $\{v_1, \dots, v_n\}$ for V_1, and set $w_i = \sigma v_i$, all i, to get a basis for V_2. Then

$$\widehat{B}_1 = [B_1(v_i, v_j)] = [B_2(\sigma v_i, \sigma v_j)] = [B_2(w_i, w_j)] = \widehat{B}_2.$$

$\Leftarrow$: Suppose $\widehat{B}_1 = \widehat{B}_2$ relative to bases $\{v_i\}$ and $\{w_i\}$ for V_1 and V_2. Define $\sigma: V_1 \to V_2$ via $\sigma(v_i) = w_i$, all i, an isomorphism. If $u, w \in V_1$ write $u = \sum_i a_i v_i$ and $w = \sum_i b_i v_i$. Then

$$B_1(u, w) = \sum_{i,j} a_i b_j B_1(v_i, v_j) = \sum_{i,j} a_i b_j B_2(w_i, w_j)$$

$$= \sum_{i,j} a_i b_j B_2(\sigma v_i, \sigma v_j) = B_2(\sum_i a_i \sigma v_i, \sum_j b_j \sigma v_j)$$
$$= B_2(\sigma u, \sigma w).$$

□

Proposition 2.9. *Suppose that B is a reflexive bilinear form on V, and that W is a nondegenerate subspace of V. Then $V = W \oplus W^{\perp}$.*

Proof. We may choose a basis $\{v_1, \ldots, v_k\}$ for W and adjoin further vectors $\{v_{k+1}, \ldots, v_n\}$ to extend it to a basis for V. Represent B by its matrix $\widehat{B} = [b_{ij}]$ relative to $\{v_i\}$, and represent vectors in V correspondingly as column vectors in F^n. If $v = \sum_i a_i v_i \in V$, so $v \mapsto (a_1, \ldots, a_n)^t$, then $v \in W^{\perp}$ if and only if $0 = B(v_i, v) = \sum_j B(v_i, v_j) a_j = \sum_j b_{ij} a_j$ for $1 \leq i \leq k$, i.e. if and only if $(a_1, \ldots, a_n)^t$ is in the solution space of the homogeneous system $AX = 0$, whose coefficient matrix A is the $k \times n$ matrix over F whose rows are the first k rows of $\widehat{B}$. Thus dim $W^{\perp} = n -$ rank $A \geq n - k$. But W nondegenerate means that rad $W = W \cap W^{\perp} = 0$, so $W + W^{\perp} = W \oplus W^{\perp}$, and $\dim(W \oplus W^{\perp}) = \dim W + \dim W^{\perp} \geq k + (n-k) = n$; hence $W \oplus W^{\perp} = V$. □

Assume for the remainder of the chapter that B is an alternate bilinear form on V.

For a simple example take V of dimension 2 and $\widehat{B} = \begin{bmatrix} 0 & 1 \\ -1 & 0 \end{bmatrix}$. Thus, if $u = \begin{bmatrix} x_1 \\ y_1 \end{bmatrix}$ and $v = \begin{bmatrix} x_2 \\ y_2 \end{bmatrix}$, then $B(u, v) = x_1 y_2 - x_2 y_1$. We'll see that this example is surprisingly general.

If $u, v \in V$ and $B(u, v) \neq 0$ then $\{u, v\}$ is linearly independent, for if $a \in F^{\star}$ then $B(u, au) = aB(u, u) = 0$. If $B(u, v) = b \neq 0$ set $u_1 = b^{-1}u$ and $v_1 = v$. Then $B(u_1, v_1) = 1$. The subspace W spanned by u_1 and v_1 is called a *hyperbolic plane*, with *hyperbolic pair* (u_1, v_1) as basis. Note that the restriction of B to W has representing matrix $\begin{bmatrix} 0 & 1 \\ -1 & 0 \end{bmatrix}$ relative to a hyperbolic pair as basis.

A bit of notation: If U and W are subspaces of V, with $U \cap W = 0$ and also $U \perp W$, we will denote their sum (which is a direct sum) by $U \oplus V$.

Theorem 2.10. *If B is an alternating form on V then*

$$V = W_1 \oplus W_2 \oplus \cdots \oplus W_r \oplus \text{rad } V,$$

a direct sum of mutually orthogonal subspaces, with each W_i a hyperbolic plane. Consequently V has a basis

$$\{u_1, v_1, \ldots, u_r, v_r, w_1, \ldots, w_{n-2r}\}$$

relative to which the representing matrix $\widehat{B}$ has block diagonal form

$$\begin{bmatrix} M & & 0 \\ & \ddots & \\ & M & \\ 0 & & 0_{n-2r} \end{bmatrix}, \quad \text{where } M = \begin{bmatrix} 0 & 1 \\ -1 & 0 \end{bmatrix}.$$

Proof. If $B = 0$ then rad $V = V$ and there is nothing to prove. Otherwise, choose a hyperbolic pair (u_1, v_1), as above, and let W_1 be the hyperbolic plane that they span. Then W_1 is a nondegenerate subspace (the determinant of the restriction of B to W_1 is 1), so $V = W_1 \oplus W_1^{\perp}$ by Proposition 2.9. Also rad $V = V^{\perp} = (W_1 \oplus W_1^{\perp})^{\perp} = W_1^{\perp} \cap W_1^{\perp\perp} = \text{rad}(W_1^{\perp})$. The result follows by induction on the dimension. □

Corollary 2.11. *B has even rank; if B is nondegenerate then* dim V *is even.*

Corollary 2.12. *Alternate bilinear forms B_1 and B_2 on spaces V_1 and V_2 are equivalent if and only if* dim V_1 = dim V_2 *and* rank B_1 = rank B_2.

Proof. Proposition 2.8. □

Corollary 2.13. *The determinant of any representing matrix for an alternating form is a square in F.*

Proof. If $\widehat{B}$ is as in the theorem then its determinant is 0 or 1; relative to another basis the matrix takes the form $D^t \widehat{B} D$, with determinant $(\det D)^2 \det \widehat{B}$. □

Chapter 3

Symplectic Groups

The symplectic groups discussed in this chapter were earlier called *abelian* groups or *abelian linear* groups (e.g. in [**46**] or [**18**]). This terminology was recognized even at the time as being ambiguous, since *abelian* was also interpreted as meaning *commutative*, whereas the symplectic groups are rarely commutative.

Suppose that V is a vector space of dimension $n = 2m > 0$ over a field F, and let B be a nondegenerate alternate form on V, as discussed in Chapter 2. An ordered basis $\{u_1, v_1, \dots, u_m, v_m\}$ for V built up of hyperbolic pairs $\{u_i, v_i\}$, as in Theorem 2.10, is called a *symplectic* basis for V, V is called a *symplectic space*, and somewhat loosely speaking the study of (V, B) is called *symplectic geometry*.

An invertible linear transformation τ of V is said to be *symplectic* if

$$B(\tau v, \tau w) = B(v, w)$$

for all $v, w \in V$, *i.e.* if τ is an isometry of V. The set of all symplectic transformations of V is evidently a subgroup of $\mathrm{GL}(V)$; it is called the *symplectic group* on V and denoted $\mathrm{Sp}(V)$. The dependence of $\mathrm{Sp}(V)$ on the form B is not significant. Since any other nondegenerate alternate form B_1 on V is equivalent with B (Corollary 2.12), the resulting symplectic group (relative to B_1) is conjugate to $\mathrm{Sp}(V)$ in $\mathrm{GL}(V)$. In other words, relative to appropriately chosen bases the two groups are represented by the *same* matrices in $\mathrm{GL}(n, F)$.

Suppose, then, that a symplectic basis has been chosen for V. If $\tau \in \mathrm{GL}(V)$ say that τ is represented, relative to that basis, by the matrix T. An easy calculation (as on page 14) shows that $\tau \in \mathrm{Sp}(V)$ if and only if

$T^t\widehat{B}T = \widehat{B}$. The resulting group of matrices is denoted by $\mathrm{Sp}(n, F)$, or also by $\mathrm{Sp}(n, q)$ if $|F| = q$, finite.

Proposition 3.1. *If* $\dim V = 2$ *then* $\mathrm{Sp}(V) = \mathrm{SL}(V)$.

Proof. We may calculate in $\mathrm{Sp}(2, F)$ via a symplectic basis, hence

$$\widehat{B} = M = \begin{bmatrix} 0 & 1 \\ -1 & 0 \end{bmatrix}.$$

By the remarks above $T = \begin{bmatrix} a & b \\ c & d \end{bmatrix} \in \mathrm{GL}(2, F)$ is in $\mathrm{Sp}(2, F)$ if and only if $T^tMT = M$, i.e.

$$\begin{bmatrix} 0 & ad - bc \\ -ad + bc & 0 \end{bmatrix} = \begin{bmatrix} 0 & 1 \\ -1 & 0 \end{bmatrix}.$$

□

Exercise

Let $\{u_1, v_1, \ldots, u_m, v_m\}$ be a symplectic basis for V, but re-order the vectors as $\{u_1, u_2, \ldots, u_m, v_1, v_2, \ldots, v_m\}$. Determine the resulting representing matrix for B. Show that every

$$\begin{bmatrix} A & 0 \\ 0 & (A^t)^{-1} \end{bmatrix}, \quad A \in \mathrm{GL}(m, F),$$

represents a symplectic transformation relative to this basis, and conclude that $\mathrm{Sp}(2m, F)$ has a subgroup isomorphic to $\mathrm{GL}(m, F)$.

Recall that if $1 \neq \tau \in \mathrm{SL}(V)$, then τ is called a *transvection*, with fixed hyperplane W, if $\tau|_W = 1_W$ and $\tau v - v \in W$ for all $v \in V$.

Let τ be a transvection with fixed hyperplane W, and suppose that $\tau \in \mathrm{Sp}(V)$. By Proposition 2.4 $\dim W^\perp = 1$, so say that $W^\perp = \langle u \rangle$. By Corollary 2.5 we have $u^\perp = W$, but also $u \in u^\perp = (W^\perp)^\perp = W$. Choose $x \in V \setminus W$, so $V = \langle x \rangle \oplus W$. Then $0 \neq \tau x - x = z$, say, in W. For any $v \in V$ write $v = bx + w$, $b \in F$ and $w \in W$, and define $f \in V^*$ via $f(v) = b$; thus $\mathrm{Ker}\ f = W$. By Corollary 2.2 there exists $y \in V$ such that $f(v) = B(v, y)$, all $v \in V$. Note that

$$\begin{aligned} \tau v &= \tau(bx + w) = b\tau x + w \\ &= b(x + z) + w = v + bz \\ &= v + f(v)z = v + B(v, y)z, \end{aligned}$$

all $v \in V$. In particular if $w \in W$ then $w = \tau w = w + B(w, y)z$, so $B(w, y) = 0$, all w, and $y \in W^\perp = \langle u \rangle$. Say that $y = cu$, $c \in F^\star$. Also, since $\tau \in \mathrm{Sp}(V)$, $B(w, x) = B(\tau w, \tau x) = B(w, x + z) = B(w, x) + B(w, z)$, so $B(w,z) = 0$ for all $w \in W$, and $z \in W^\perp$, say $z = du$, $d \in F^\star$. Thus $\tau v = v + B(v, cu) \cdot du = v + cdB(v,u)u$. Set $a = cd$ and we have

$$\tau v = v + aB(v,u)u,$$

all $v \in V$, i.e. τ is determined by the scalar a and the vector u. Write $\tau = \tau_{u,a}$, and call τ a *symplectic transvection* with direction u.

Conversely, given any nonzero $u \in V$ and any $a \in F^\star$ we may define $\tau = \tau_{u,a}$ via $\tau v = v + aB(v,u)u$, all v. Then clearly τ is a transvection with fixed hyperplane $W = u^\perp$, and it is easy to check directly that $B(\tau v, \tau w) = B(v,w)$, all $v, w \in V$, so $\tau_{u,a} \in \mathrm{Sp}(V)$.

It will be convenient to extend the notation slightly and define $\tau_{u,a}$ by the formula above even when $a = 0$. Of course $\tau_{u,0}$ is simply the identity transformation.

Exercise

Fix $u \neq 0$ in V.

1. Show that $\tau_{u,a}\tau_{u,b} = \tau_{u,a+b}$, $\tau_{bu,a} = \tau_{u,ab^2}$, and $\tau_{u,a}^{-1} = \tau_{u,-a}$.
2. Conclude that $\{\tau_{u,a} \colon a \in F\}$ is a group isomorphic with the additive group F.
3. If $\sigma \in \mathrm{Sp}(V)$ then ${}^\sigma\tau_{u,a} = \tau_{\sigma u,a}$.

If V is a symplectic space write $\mathcal{T} = \langle \tau_{u,a} \colon 0 \neq u \in V,\ a \in F^\star \rangle$, the subgroup of $\mathrm{Sp}(V)$ generated by symplectic transvections. It will be shown eventually that $\mathcal{T} = \mathrm{Sp}(V)$, so $\mathrm{Sp}(V)$ is generated by symplectic transvections.

Proposition 3.2. *If V is symplectic then the group $\mathcal{T} \leq \mathrm{Sp}(V)$ is transitive on $V \setminus \{0\}$.*

Proof. Take $v \neq w$ in $V \setminus \{0\}$. If $B(v,w) \neq 0$ set $a = 1/B(v,w)$ and $u = v - w$. Then

$$\tau_{u,a}(v) = v + aB(v,u)u = v + \frac{B(v, v-w)}{B(v,w)}(v-w) = v - (v-w) = w.$$

Suppose then that $B(v,w) = 0$. Choose $f \in V^*$ with both $f(v)$ and $f(w)$ nonzero. By Corollary 2.2 there exists $u \in V$ such that $f(v) = B(v,u) \neq 0$ and $f(w) = B(w,u) \neq 0$. By the first part of the proof there exist τ_1, $\tau_2 \in \mathcal{T}$ with $\tau_1 v = u$ and $\tau_2 u = w$, hence $\tau_2\tau_1 v = w$. □

Recall from Chapter 2 that a *hyperbolic pair* in V is an ordered pair (u,v) of vectors (necessarily linearly independent) with $B(u,v) = 1$.

Proposition 3.3. *If V is symplectic then the group $\mathcal{T} \leq \mathrm{Sp}(V)$ is transitive on the set $\mathcal{S}$ of all hyperbolic pairs in V.*

Proof. Take (u_1, v_1), $(u_2, v_2) \in \mathcal{S}$. First use Proposition 3.2 to choose $\tau \in \mathcal{T}$ with $\tau u_1 = u_2$, and hence $\tau\colon (u_1, v_1) \mapsto (u_2, v_3) \in \mathcal{S}$, $v_3 = \tau v_1$. Now we need $\sigma \in \mathcal{T}$ with $\sigma\colon (u_2, v_3) \mapsto (u_2, v_2)$, i.e. $\sigma \in \mathrm{Stab}_{\mathcal{T}}(u_2)$ and $\sigma v_3 = v_2$. If $B(v_3, v_2) \neq 0$ use $\sigma = \tau_{u,a}$, with $u = v_3 - v_2$, as in the proof of Proposition 3.2. Then $\sigma v_3 = v_2$ and also $B(u_2, v_3) = 1 = B(u_2, v_2)$, so $B(u_2, u) = B(u_2, v_3 - v_2) = 0$, and therefore $\sigma u_2 = u_2 + aB(u_2, u)u = u_2$.

Suppose then that $B(v_3, v_2) = 0$. Note that $(u_2, u_2 + v_3) \in \mathcal{S}$, and that $B(v_3, u_2 + v_3) \neq 0$, so by what was just proved there exists $\sigma_1 \in \mathcal{T}$ with $\sigma_1 u_2 = u_2$ and $\sigma_1 v_3 = u_2 + v_3$. But also $B(u_2 + v_3, v_2) = B(u_2, v_2) \neq 0$, so there exists $\sigma_2 \in \mathcal{T}$ with $\sigma_2 u_2 = u_2$ and $\sigma_2(u_2 + v_3) = v_2$. Setting $\sigma = \sigma_2\sigma_1 \in \mathcal{T}$ we have $\sigma u_2 = u_2$ and $\sigma v_3 = \sigma_2(u_2 + v_3) = v_2$. Finally, $\sigma\tau \in \mathcal{T}$ and $\sigma\tau(u_1, v_1) = (u_2, v_2)$. □

Theorem 3.4. *The symplectic group* $\mathrm{Sp}(V)$ *is generated by transvections.*

Proof. Use induction on m, where $2m = n = \dim V$. The case $m = 1$ is covered by Proposition 3.1 and Theorem 1.4.

Choose a hyperbolic pair (u, v) in V and set $W = \langle u, v\rangle$, a hyperbolic plane. Then $V = W \perp W^\perp$ by Proposition 2.9. Take any $\sigma \in \mathrm{Sp}(V)$. Then $(\sigma u, \sigma v)$ is a hyperbolic pair, so by Proposition 3.3 there exists $\tau \in \mathcal{T}$ with $\tau\sigma u = u$ and $\tau\sigma v = v$, so $\tau\sigma|_W = 1_W$. But also $\tau\sigma|_{W^\perp} \in \mathrm{Sp}(W^\perp)$. By induction $\tau\sigma|_{W^\perp}$ is a product of symplectic transvections on $W^\perp$. Any such transvection, say $\tau_{w,a}$, determines, by the *same* formula, viz. $\tau_{w,a}(v) = v + aB(v, w)w$, a symplectic transvection on V whose fixed hyperplane contains W. Since also $\tau\sigma|_W = 1_W$ we conclude that $\tau\sigma \in \mathcal{T}$, and hence that $\sigma \in \mathcal{T}$. □

Corollary 3.5. $\mathrm{Sp}(V) \leq \mathrm{SL}(V)$.

Proposition 3.6. *The center of* $\mathrm{Sp}(V)$ *is* $\{\pm 1\}$.

Proof. Note that $\tau_{u,a} v = v$ if and only if $B(v, u) = 0$, so $u^\perp$ is the space of fixed points for $\tau_{u,a}$. If $\sigma \in Z(\mathrm{Sp}(V))$ then in particular σ commutes with every $\tau_{u,a}$. If $v \in u^\perp$ then $\tau_{u,a}(\sigma v) = \sigma\tau_{u,a}v = \sigma v$, so $\sigma v \in u^\perp$, i.e. $\sigma u^\perp = u^\perp$ for all u. But $\langle u\rangle = \mathrm{rad}\, u^\perp$ and σ is symplectic, so σ carries $\langle u\rangle$ to $\langle u\rangle$; in particular, $\sigma u = c_u u$ for some $c_u \in F$, all $u \in V$. Choose u and v linearly independent in V. Then $\sigma(u + v) = c_{u+v}(u + v) = c_u u + c_v v$, so $c_u = c_{u+v} = c_v = c$, say, independent of u and v, and $\sigma = c 1_V$. But then

$$B(u, v) = B(\sigma u, \sigma v) = c^2 B(u, v)$$

for all u, v, so $c = \pm 1$. □

Proposition 3.7. *If* $|F| \geq 4$ *then the derived group of* $\mathrm{Sp}(V)$ *is* $\mathrm{Sp}(V)$.

Proof. Fix $u \neq 0$ in V and a in $F^\star$. Choose $b \in F \setminus \{0, \pm 1\}$ and set

$$c = a/(1 - b^2), \quad d = -b^2 c.$$

Then $c + d = a$, so $\tau_{u,c}\tau_{u,d} = \tau_{u,a}$ by the exercise on page 23. Choose $\sigma \in \mathrm{Sp}(V)$ with $\sigma u = bu$ (Proposition 3.2). Then

$$\sigma\tau_{u,c}^{-1}\sigma^{-1} = \sigma\tau_{u,-c}\sigma^{-1} = \tau_{\sigma u,-c} = \tau_{bu,-c} = \tau_{u,-b^2 c} = \tau_{u,d},$$

and hence

$$\tau_{u,c}\sigma\tau_{u,c}^{-1}\sigma^{-1} = \tau_{u,c}\tau_{u,d} = \tau_{u,a} \in \mathrm{Sp}'(V).$$

Apply Theorem 3.4. □

Proposition 3.8. *If* $|F| = 3$ *and* $\dim V \geq 4$ *then the derived group of* $\mathrm{Sp}(V)$ *is* $\mathrm{Sp}(V)$.

Proof. Choose a symplectic basis $\{u_1, v_1, \ldots, u_m, v_m\}$ and define linear transformations σ and τ as follows:

$$\sigma\colon u_1 \mapsto u_1 + u_2,\ v_1 \mapsto v_2,\ u_2 \mapsto u_1,\ v_2 \mapsto v_1 - v_2,$$

and

$$u_i \mapsto u_i,\ v_i \mapsto v_i \ \text{ if } i > 2;$$

$$\tau\colon u_1 \mapsto u_1 - v_1 + v_2,\ v_1 \mapsto v_1,\ u_2 \mapsto u_2 + v_1,\ v_2 \mapsto v_2,$$

and

$$u_i \mapsto u_i,\ v_i \mapsto v_i \ \text{ if } i > 2.$$

It is an easy exercise to show that σ and τ are in $\mathrm{Sp}(V)$ and to show that $\sigma\tau\sigma^{-1}\tau^{-1} = \tau_{v_1,1}$. Thus $\tau_{v_1,1} \in \mathrm{Sp}'(V)$. Given any $u \neq 0$ in V, choose $\gamma \in \mathrm{Sp}(V)$ with $\gamma v_1 = u$; then

$$\gamma\tau_{v_1,1}\gamma^{-1} = \tau_{\gamma v_1,1} = \tau_{u,1} \in \mathrm{Sp}'(V)$$

and also

$$\tau_{u,1}^{-1} = \tau_{u,-1} \in \mathrm{Sp}'(V).$$

Apply Theorem 3.4. □

The case $|F| = 3$ and $\dim V = 2$ is an exception, since $\mathrm{Sp}(2,3) = \mathrm{SL}(2,3)$ of order 24.

Proposition 3.9. *If* $|F| = 2$ *and* $\dim V \geq 6$ *then the derived group of* $\mathrm{Sp}(V)$ *is* $\mathrm{Sp}(V)$.

Proof. The proof is exactly as for Proposition 3.8, except that

$$\sigma\colon u_1 \mapsto u_1 + u_3,\ v_1 \mapsto v_3,\ u_2 \mapsto u_1,\ v_2 \mapsto v_1 + v_3,$$

$$u_3 \mapsto u_2,\ v_3 \mapsto v_2,$$

and

$$u_i \mapsto u_i,\ v_i \mapsto v_i \ \text{ if } i > 3;$$

$$\tau\colon u_1 \mapsto u_1 + u_2,\ v_1 \mapsto v_1,\ u_2 \mapsto v_1 + u_2 + v_2 + v_3,\ v_2 \mapsto v_2,$$

$$u_3 \mapsto v_2 + u_3 + v_3,\ v_3 \mapsto v_3,$$

and

$$u_i \mapsto u_i,\ v_i \mapsto v_i \ \text{ if } i > 3.$$

□

Since $\mathrm{Sp}(2,2) = \mathrm{SL}(2,2) \cong \mathrm{Sym}(3)$ it is an exception to Proposition 3.9. It will be shown later (Proposition 3.13) that $\mathrm{Sp}(4,2) \cong \mathrm{Sym}(6)$, so it is also an exception.

Since $\mathrm{Sp}(V) \leq \mathrm{GL}(V)$ there is an action of $\mathrm{Sp}(V)$ on the projective space $\mathbb{P} = \mathbb{P}_{n-1}(V)$. The kernel of the action is $Z(\mathrm{Sp}(V)) = \{\pm 1\}$, and we define the *projective symplectic group* $\mathrm{PSp}(V)$ to be $\mathrm{Sp}(V)/Z(\mathrm{Sp}(V))$; it acts faithfully and transitively (see Proposition 3.2) on $\mathbb{P}$.

Proposition 3.10. $\mathrm{Sp}(V)$ *is primitive on* $\mathbb{P} = \mathbb{P}_{n-1}(V)$.

Proof. We may assume that $n \geq 4$ by Proposition 3.1, since $\mathrm{SL}(2,F)$ is doubly transitive. Suppose that $S \subseteq \mathbb{P}$, $|S| > 1$, and either $\sigma S = S$ or $\sigma S \cap S = \emptyset$ for each $\sigma \in \mathrm{Sp}(V)$. We show first that there are $[u]$, $[v] \in S$ with $B(u,v) \neq 0$. Suppose to the contrary that $B(u,v) = 0$ for all $[u]$, $[v] \in S$. Choose $[u] \neq [v]$ in S and choose $f \in V^*$ with $f(u) = 1$, $f(v) = 0$. By Corollary 2.2 there exists $x \in V$ with $B(u,x) = 1$, $B(v,x) = 0$. Set $W = \mathrm{Span}(u,x)$, a hyperbolic plane, and set $H = \{\sigma \in \mathrm{Sp}(V)\colon \sigma|_W = 1_W\} \leq \mathrm{Sp}(V)$. Since $V = W \oplus W^{\perp}$ it is clear that every $\sigma \in \mathrm{Sp}(W^{\perp})$ extends trivially to $\sigma' \in \mathrm{Sp}(V)$ with $\sigma'|_W = 1_W$, so in fact $\mathrm{Sp}(W^{\perp}) = \{\tau|_{W^{\perp}}\colon \tau \in H\}$. Choose $w \neq 0$ in $W^{\perp}$. Since $v \in W^{\perp}$ there is some $\tau \in H$ with $\tau v = w$, by Proposition 3.2. Since $\tau u = u$ we have $[u] \in S \cap \tau S$, so $\tau S = S$. Also $[w] = \tau[v] \in S$, and w was an arbitrary nonzero vector in $W^{\perp}$ Since $W^{\perp} \neq 0$ there is a hyperbolic pair (y,z) in $W^{\perp}$; hence $[y]$, $[z] \in S$, but $B(y,z) = 1$, a contradiction.

Thus we may in fact choose $[u]$, $[v] \in S$ with $B(u,v) \neq 0$, so we may assume that $B(u,v) = 1$ and (u,v) is a hyperbolic pair. Now take any $[w] \in \mathbb{P}$. If $B(u,w) \neq 0$ we may assume that (u,w) is a hyperbolic pair and by Proposition 3.3 find $\sigma \in \mathrm{Sp}(V)$ with $\sigma u = u$ and $\sigma v = w$. Thus

$u \in S \cap \sigma S$, so $\sigma S = S$ and $[w] \in S$. On the other hand, if $B(u,w) = 0$ choose $f \in V^*$ with $f(u) = f(w) = 1$ and thereby obtain $x \in V$ with $B(u,x) = B(w,x) = 1$. As above, $[x] \in S$, and also there exists $\tau \in \mathrm{Sp}(V)$ with $\tau u = w$ and $\tau x = x$. Again $\tau S = S$, since $[x] \in S \cap \tau S$, and $[w] = \tau[u] \in S$. Thus $S = \mathbb{P}$. □

Theorem 3.11. *Except for* PSp(2, 2), PSp(2, 3), *and* PSp(4, 2) *every projective symplectic group* PSp(V) *is simple.*

Proof. We know that PSp(V) acts faithfully on $\mathbb{P} = \mathbb{P}_{n-1}(V)$, and the action is primitive by Proposition 3.10. With the exceptions noted, PSp(V) is its own derived group by Propositions 3.7, 3.8, and 3.9. Fix $[u] \in \mathbb{P}$ and set $H = \mathrm{Stab}_{\mathrm{Sp}(V)}([u])$, $\overline{H} = H/\{\pm 1\} = \mathrm{Stab}_{\mathrm{PSp}(V)}([u])$. Set $K = \{\tau_{u,a} \colon a \in F\}$. By the exercise on page 23 $K \lhd H$ and $K \cong F$, so K is abelian. If $\sigma \in \mathrm{Sp}(V)$ then ${}^{\sigma}K \supseteq \{\tau_{\sigma u,a} \colon a \in F\}$, so $\cup\{{}^{\sigma}K \colon \sigma \in \mathrm{Sp}(V)\}$ contains all symplectic transvections by Proposition 3.2, and hence generates Sp(V) by Theorem 3.4. Thus $\langle {}^{\overline{\sigma}}\overline{K} \colon \overline{\sigma} \in \mathrm{PSp}(V)\rangle = \mathrm{PSp}(V)$, and with the exceptions noted PSp(V) is simple by Iwasawa's Theorem (0.5). □

PSp($2n, F$) is the Chevalley group $C_n(F)$ (see Carter [**11**], page 184).

Let us calculate the orders of the finite symplectic groups.

Theorem 3.12. *If* $|F| = q$, *finite, then*

$$|\mathrm{Sp}(n,q)| = q^{m^2}\prod_{i=1}^{m}(q^{2i}-1),$$

where $n = 2m$, *and*

$$|\mathrm{PSp}(n,q)| = \frac{|\mathrm{Sp}(n,q)|}{(2,q-1)}.$$

Proof. Any one of the $q^n - 1$ nonzero vectors can serve as the first vector u in a hyperbolic pair. Since $|q^{\perp}| = q^{n-1}$ there are $q^n - q^{n-1}$ vectors v with $B(u,v) \neq 0$. Thus there are $(q^n - q^{n-1})/(q-1) = q^{n-1}$ choices of v for which $B(u,v) = 1$, and consequently $(q^n - 1)q^{n-1}$ distinct hyperbolic pairs. If (u,v) is a hyperbolic pair and W is the hyperbolic plane with $\{u,v\}$ as basis, then $\sigma \in \mathrm{Stab}_{\mathrm{Sp}(V)}((u,v))$ if and only if $\sigma|_W = 1_W$, and we saw in the proof of Proposition 3.10 that $\mathrm{Stab}_{\mathrm{Sp}(V)}((u,v)) \cong \mathrm{Sp}(W^{\perp})$. Thus

$$|\mathrm{Sp}(n,q)| = (q^n - 1)q^{n-1}|\mathrm{Sp}(n-2,q)|$$

by Proposition 0.1, and we may argue by induction. When $n = 2$ then $|\mathrm{Sp}(2,q)| = |\mathrm{SL}(2,q)| = q(q^2-1)$. Assume inductively that

$$|\mathrm{Sp}(n-2,q)| = q^{(m-1)^2}\prod_{i=1}^{m-1}(q^{2i}-1),$$

then

$$|\operatorname{Sp}(n,q)| = q^{2m-1}(q^{2m}-1)q^{(m-1)^2}\prod_{i=1}^{m-1}(q^{2i}-1) = q^{m^2}\prod_{1}^{m}(q^{2i}-1).$$

The relation to $|\operatorname{PSp}(n,q)|$ is immediate from Proposition 3.6. □

Note, for example, that $|\operatorname{Sp}(4,2)| = 720 = |\operatorname{Sym}(6)|$. Let us prepare to show that $\operatorname{Sp}(4,2) \cong \operatorname{Sym}(6)$.

If V has dimension 4 over $F = \mathbb{F}_2$ then $|V| = 16$. Choose $v_1 \neq 0$ in V, and then choose $v_2 \in V \setminus v_1^\perp$. Then $v_2 \in v_2^\perp \setminus v_1^\perp$, so $v_1^\perp + v_2^\perp = V$, and therefore $\dim(v_1^\perp \cap v_2^\perp) = 3 + 3 - 4 = 2$ and $|v_1^\perp \cap v_2^\perp| = 4$. Thus $|v_1^\perp \cup v_2^\perp| = 8 + 8 - 4 = 12$, and we may choose $v_3 \in V \setminus \{v_1^\perp \cup v_2^\perp\}$. Similar inclusion-exclusion arguments show that $|v_1^\perp \cup v_2^\perp \cup v_3^\perp| = 14$, so there exists $v_4 \in V \setminus \{v_1^\perp \cup v_2^\perp \cup v_3^\perp\}$. Then $|v_1^\perp \cup v_2^\perp \cup v_3^\perp \cup v_4^\perp| = 15$, and the last remaining element v_5 of V is in $V \setminus \{v_1^\perp \cup v_2^\perp \cup v_3^\perp \cup v_4^\perp\}$. The resulting set $\{v_1, \dots, v_5\}$ satisfies $B(v_i, v_j) = 1$ if $i \neq j$, and is a maximal nonorthogonal set in V.

Now we approach maximal nonorthogonal sets from another direction. Choose a hyperbolic pair (v_1, v_2) in V and set $W = \langle v_1, v_2\rangle$. Then $V = W \oplus W^\perp$ and we may choose a hyperbolic pair (x_1, x_2) in $W^\perp$. If we set $v_3 = v_1 + v_2 + x_1$, $v_4 = v_1 + v_2 + x_2$, and $v_5 = v_1 + v_2 + x_1 + x_2$, then it is a routine matter to check that $\{v_1, \dots, v_5\}$ is a maximal nonorthogonal set (m.n.s.).

It is in fact the *unique* m.n.s. containing $\{v_1, v_2\}$. The reason: given (v_1, v_2), suppose that we try to adjoin w so that $\{v_1, v_2, w\}$ is a nonorthogonal set. Write $w = a_1v_1 + a_2v_2 + b_1x_1 + b_2x_2$. Then $1 = B(v_1, w) = a_2$ and $1 = B(v_2, w) = a_1$, and $w = v_1 + v_2 + b_1x_1 + b_2x_2$. If we try taking $b_1 = b_2 = 0$, giving, say, $w' = v_1 + v_2$, then $w' \perp w$ for any other choice of values of b_1 and b_2, so $\{v_1, v_2, w'\}$ is not contained in a larger nonorthogonal set. There remain only the other three possible assignments of values to b_1 and b_2.

By the proof of Theorem 3.12 there are $(2^4 - 1)\cdot 2^{4-1} = 120$ hyperbolic pairs in V. Each m.n.s. contains $5 \cdot 4 = 20$ hyperbolic pairs, so there are exactly $120/20 = 6$ distinct m.n.s.'s in V.

Proposition 3.13. $\operatorname{Sp}(4,2) \cong \operatorname{Sym}(6)$.

Proof. Let $\mathcal{S}$ be the set of 6 m.n.s.'s in V, as above. It follows from Proposition 3.3 that $\operatorname{Sp}(4,2)$ acts transitively on $\mathcal{S}$, so there is a homomorphism $\varphi\colon \operatorname{Sp}(4,2) \to \operatorname{Sym}(6)$. Let $K = \operatorname{Ker}\varphi$. Choose a hyperbolic pair (v_1, v_2), set $W = \langle v_1, v_2\rangle$, and choose a hyperbolic pair (x_1, x_2) in $W^\perp$. As above we obtain two m.n.s.'s $\{v_i\}$ and $\{x_i\}$, with

$$\begin{aligned} v_3 &= v_1+v_2+x_1, & v_4 &= v_1+v_2+x_2, & v_5 &= v_1+v_2+x_1+x_2; \\ x_3 &= x_1+x_2+v_1, & x_4 &= x_1+x_2+v_2, & x_5 &= x_1+x_2+v_1+v_2. \end{aligned}$$

Set $z = v_1+v_2+x_1+x_2$ and note that $\{v_i\}\cap\{x_i\} = \{z\}$. Every $\sigma \in K$ fixes $\{v_i\}$ and $\{x_i\}$, so σ fixes $\{v_i\}\cap\{x_i\}$; i.e. $\sigma\, z = z$, all $\sigma \in K$. But $\mathrm{Sp}(V)$ is transitive on $P(V) = V\setminus\{0\}$, and for each $\tau \in \mathrm{Sp}(V)$ we have ${}^{\tau}\sigma(\tau\, z) = \tau\, z$; i.e. $\tau\, z$ is fixed by all ${}^{\tau}\sigma \in K$, which is all of K. Since $\mathrm{Sp}(V) = \mathrm{PSp}(V)$ acts faithfully on $P(V)$ we conclude that $K = 1$. Thus φ is 1–1, and hence is an isomorphism since $|\,\mathrm{Sp}(V)| = |\,\mathrm{Sym}(6)|$. □

Thus $\mathrm{Sp}(4,2)$ is not simple, but its derived group ($\cong \mathrm{Alt}(6)$) is simple and of index 2.

Exercises

1. Suppose that V is a space with an alternate bilinear form B and that W is a subspace. Set $U = W/\operatorname{rad} W$. Show that U is a symplectic space if we define
$$\overline{B}(w_1 + \operatorname{rad} W, w_2 + \operatorname{rad} W) = B(w_1, w_2).$$
2. (This is an alternate approach to Proposition 3.13.) Take V of dimension 6 over $F = \mathbb{F}_2$ with basis $\{v_1, \ldots, v_6\}$. Define B via $B(v_i, v_j) = 1 + \delta_{ij}$, and verify that (V, B) is a symplectic space. Observe that $\mathrm{Sym}(6)$ is embedded isomorphically in $\mathrm{Sp}(V)$ by permuting the elements of $\{v_1, \ldots, v_6\}$. Set
$$W = \langle v_1 + \cdots + v_6\rangle;$$
W and $W^{\perp}$ are $\mathrm{Sym}(6)$-invariant, so $\mathrm{Sym}(6)$ acts on $U = W^{\perp}/W$, a 4-dimensional symplectic space by Exercise 1 above. The transposition $(1\,2) \in \mathrm{Sym}(6)$ embeds as the transvection $\tau_{v_1+v_2,1}$ in $\mathrm{Sp}(V)$ and maps to a transvection in $\mathrm{Sp}(U)$. Its conjugates in $\mathrm{Sym}(6)$ yield all 15 transvections in $\mathrm{Sp}(U)$, and therefore $\mathrm{Sym}(6) \cong \mathrm{Sp}(U)$. Provide the details.

Note that $|\,\mathrm{PSp}(4,3)| = 3^4\cdot(3^2-1)(3^4-1)/2 = 25{,}920$. A. Cayley and G. Salmon proved in the 1840s that each sufficiently general cubic surface has 27 straight lines lying on it. C. Jordan (who first defined the symplectic groups) determined the group of incidence-preserving permutations of the 27 lines, and found it to have order 51,840. The group is in fact the Weyl group of type E_6 (e.g. see [**29**] or [**40**]), and it has a copy of $\mathrm{PSp}(4,3)$ as a subgroup of index 2.

In classical plane geometry it was known that each sufficiently general quartic curve has 28 "bitangents," i.e. lines that are tangent to the curve at two distinct points. This seems to have first been proved rigorously by K. Jacobi in 1850. Jordan showed that the symmetry group of the 28 bitangents is a simple group of order 1,451,520 – it is isomorphic with $\mathrm{PSp}(6,2)$.

For further discussion of the history of these configurations and relations between them, and for further references, see [**71**]. Also see the Atlas [**15**], where the groups are denoted $S_4(3)$ and $S_6(2)$.

Chapter 4

Symmetric Forms and Quadratic Forms

We begin the study of a rather far-reaching generalization of the familiar Euclidean geometry of n-dimensional (especially 2- and 3-dimensional) real vector spaces. A number of subtleties will unfold. In particular, vector spaces over fields of characteristic 2 will require special treatment in later chapters.

Assume for this chapter that F is a field with $\operatorname{char} F \neq 2$.

Recall from Chapter 2 that a bilinear form B on a vector space V over F is *symmetric* if $B(u, v) = B(v, u)$ for all u, $v \in V$. Thus any representing matrix for B is a symmetric matrix.

If B is a symmetric form on V define $Q\colon V \to F$ via $Q(v) = B(v, v)$, and call Q the associated *quadratic form*. Note that

$$Q(av) = a^2 Q(v)$$

for all $a \in F$, all $v \in V$, and that

$$Q(u + v) = B(u + v, u + v) = Q(u) + 2B(u, v) + Q(v)$$

for all u, $v \in V$. Thus

$$B(u, v) = \frac{1}{2}[Q(u + v) - Q(u) - Q(v)],$$

and the bilinear form B is completely determined by the quadratic form Q.

If $\{v_1, v_2, \dots, v_n\}$ is a basis for V, and $\widehat{B} = [b_{ij}]$, as usual, and $v \in V$, write $v = \sum_i a_i v_i$. Then

$$Q(v) = B(v, v) = \sum_{i,j} b_{ij} a_i a_j,$$

so Q is a homogeneous degree 2 polynomial function in the n coordinates of V relative to the basis. That is, in fact, the usual definition of a quadratic form.

Proposition 4.1. *If B is a nonzero symmetric bilinear form on V, with quadratic form Q, then $Q(v) \neq 0$ for some $v \in V$.*

Proof. If $Q(u) = 0$ for all u, then

$$0 = Q(u + v) = Q(u) + 2B(u, v) + Q(v) = 2B(u, v),$$

so $B(u, v) = 0$ for all u, $v \in V$, a contradiction. □

A set $\{v_1, \dots, v_k\}$ of vectors in V is called *orthogonal* (relative to B) if $v_i \perp v_j$ (i.e. $B(v_i, v_j) = 0$) for $i \neq j$.

Theorem 4.2. *If B is a symmetric bilinear form on V then V has an orthogonal basis $\{v_1, \dots, v_n\}$, relative to which B has diagonal representing matrix*

$$\widehat{B} = \begin{bmatrix} b_1 & 0 & & \\ 0 & \ddots & & 0 \\ & & b_r & \\ & 0 & & 0 \end{bmatrix},$$

with all $b_i \neq 0$, $r = \operatorname{rank}(B)$, and $\{v_{r+1}, \dots, v_n\}$ a basis for $\operatorname{rad}(V)$.

Proof. We may assume $B \neq 0$ and use induction on n. Use Proposition 4.1 to choose $v_1 \in V$ with $Q(v_1) = b_1 \neq 0$, and set $W = Fv_1$. Then W is nondegenerate, so $V = W \oplus W^\perp$ by Proposition 2.9. By induction there are $v_2, \dots, v_n$, comprising an orthogonal basis for $W^\perp$ such that $Q(v_i) = b_i \neq 0$ if $2 \leq i \leq r$ and $Q(v_i) = 0$ if $r + 1 \leq i \leq n$ (for some r). Then $\widehat{B}$ is as required, so clearly $r = \operatorname{rank}(B)$. If $v \in V$ then $v \in \operatorname{rad} V$ if and only if $v \perp v_i$, all i. Write $v = \sum_i a_i v_i$; then

$$B(v, v_j) = \begin{cases} a_j b_j & \text{if } 1 \leq j \leq r, \\ 0 & \text{if } j > r. \end{cases}$$

Thus $v \in \operatorname{rad} B$ if and only if $a_j = 0$ for $1 \leq j \leq r$, hence if and only if $v \in \langle v_{r+1}, \dots, v_n \rangle$. □

Corollary 4.3. *Given a basis $\{v_1, \dots, v_n\}$ as in the theorem, then for any $v = \sum_i a_i v_i$ we have $Q(v) = \sum_i b_i a_i^2$ (all "cross terms" are eliminated).*

(Important) Note I. Any v_i in Theorem 4.2 can be replaced by $c_i v_i$ for any nonzero $c_i \in F$. The effect is to replace the entry b_i in $\widehat{B}$, $1 \leq i \leq r$, by $Q(c_i v_i) = c_i^2 b_i$.

(Important) Note II. Observe also that v_1 was chosen arbitrarily enough so that b_1 could be any element in the image of Q, i.e. any element of F "represented by" Q. At each stage b_i can be any field element represented by the restriction of Q to $\langle v_1, \dots, v_{i-1}\rangle^{\perp}$.

Theorem 4.4. *Suppose F is a field in which every element has a square root (e.g. F algebraically closed). Then symmetric bilinear forms B_1 and B_2 on spaces V_1 and V_2 of the same dimension are equivalent if and only if B_1 and B_2 have the same rank.*

Proof. By Theorem 4.2 and the **Notes** following it any such form of rank r can be represented by the matrix $\begin{bmatrix} I & 0 \\ 0 & 0 \end{bmatrix}$, where I is an $r \times r$ identity matrix. Apply Proposition 2.8. □

The next two theorems will not be used in the sequel – they are included for historical interest and to give some indication of possible further interesting research topics.

Theorem 4.5 (Sylvester's Law of Inertia). *Suppose F is an ordered field (e.g. $F \subseteq \mathbb{R}$), B is a symmetric bilinear form on a vector space V over F, and $\{u_1, \dots, u_n\}$, $\{v_1, \dots, v_n\}$ are orthogonal bases for V for which the matrices representing B are, respectively,*

$$\left[\begin{array}{ccc|ccc|c} b_1 & & 0 & & & & \\ & \ddots & & & 0 & & 0 \\ 0 & & b_p & & & & \\ \hline & & & -b_{p+1} & & 0 & \\ & 0 & & & \ddots & & 0 \\ & & & 0 & & -b_r & \\ \hline & 0 & & & 0 & & 0 \end{array}\right]$$

and

$$\left[\begin{array}{ccc|ccc|c} d_1 & & 0 & & & & \\ & \ddots & & & 0 & & 0 \\ 0 & & d_q & & & & \\ \hline & & & -d_{q+1} & & 0 & \\ & 0 & & & \ddots & & 0 \\ & & & 0 & & -d_r & \\ \hline & 0 & & & 0 & & 0 \end{array}\right],$$

with all $b_i > 0$ and all $d_i > 0$. Then $p = q$.

Proof. Set $W_1 = \langle u_1, \ldots, u_p\rangle$ and $W_2 = \langle v_{q+1}, \ldots, v_n\rangle$. If $0 \neq v \in W_1$, with $v = \sum_1^p a_i u_i$, then $Q(v) = \sum_1^p b_i a_i^2 \geq 0$. If $w \in W_2$, with $w = \sum_{q+1}^n c_i v_i$, then $Q(w) = -\sum_{q+1}^n d_i c_i^2 \leq 0$. Thus $W_1 \cap W_2 = 0$, so $\dim W_1 + \dim W_2 \leq n$, or $p + (n-q) \leq n$, or $p \leq q$. Similarly $q \leq p$. □

The point of Theorem 4.5 is that the number p of positive (or negative) entries in a diagonal representing matrix for B is independent of the chosen basis for V. Thus we may define the *signature* of B to be the number of positive diagonal entries minus the number of negative entries (hence $p - (r - p) = 2p - r$). Denote the signature by $\mathrm{sig}(B) = 2p - r$.

Note that if $\mathrm{rank}\, B = r$ and $\mathrm{sig}(B) = s$ then $(r+s)/2$ is the number of positive diagonal entries in $\widehat{B}$ and $(r-s)/2$ the number of negative entries.

The theorem is important, for example, in the classification of real conics and quadrics, which is carried out in most linear algebra textbooks.

Exercise

Take $F = \mathbb{Q}$ and $V = K = \mathbb{Q}(\sqrt[3]{2})$, viewed as a vector space over F. Define $B(u,v) = \mathrm{Tr}_{K/F}(uv)$ for u, $v \in V$. Show that B is a nondegenerate symmetric bilinear form on V, find an orthogonal basis, and determine $\mathrm{sig}(B)$.

Theorem 4.6. *If F is an ordered field in which every positive element has a square root (e.g. $F = \mathbb{R}$ or F real-closed), and if B_1, B_2 are symmetric bilinear forms on F-spaces V_1, V_2 of the same dimension, then B_1 and B_2 are equivalent if and only if* $\mathrm{rank}\, B_1 = \mathrm{rank}\, B_2$ *and* $\mathrm{sig}(B_1) = \mathrm{sig}(B_2)$.

Proof. Apply **Notes I** and **II** on page 33 to get representing matrices of the form

$$\begin{bmatrix} I_p & 0 & 0 \\ 0 & -I_{r-p} & 0 \\ 0 & 0 & 0 \end{bmatrix}$$

for B_1 and B_2. Then apply Proposition 2.8. □

The question of equivalence of quadratic forms is in general unsolved, and seems to be very difficult. There is a solution in case F is an algebraic number field (the Hasse-Minkowski Theorem). Roughly speaking, the answer in that case is that B_1 and B_2 are equivalent if and only if they are equivalent for all completions of F (see e.g. **[50]** for further details).

After some preliminaries we will be able to answer the equivalence question for finite fields.

If B is a symmetric form on V, with quadratic form Q, then $v \in V$, $v \neq 0$, is called *isotropic* if $Q(v) = 0$, *anisotropic* if $Q(v) \neq 0$. For convenience,

$v = 0$ is taken to be anisotropic, even though $Q(0) = 0$. If there exists an isotropic vector then B and V (and Q) are called *isotropic*; otherwise they are *anisotropic.*

The form B and quadratic form Q are called *universal* if $Q(V) = F$, i.e. given $a \in F$ there is some $v \in V$ with $Q(v) = a$.

Proposition 4.7. *If B is a nondegenerate isotropic symmetric bilinear form then B is universal.*

Proof. Let $u \neq 0$ be an isotropic vector in V. Since B is nondegenerate there is some $w \in V$ with $B(u,w) = b \neq 0$. By replacing w by $w/2b$ we may assume that $B(u,w) = 1/2$. Set $v = cu + w$, where $c \in F$ is yet to be determined. Then

$$Q(v) = B(cu + w, cu + w) = 2cB(u,w) + B(w,w) = c + B(w,w).$$

Now given any $a \in F$ set $c = a - B(w,w)$. Then

$$Q(v) = a - B(w,w) + B(w,w) = a.$$

□

Suppose now that F is a finite field, with $|F| = q$ (odd). The squaring map $\theta\colon F^\star \to F^{\times 2}$, defined by $\theta(a) = a^2$, $a \in F^\star$, is a homomorphism, and $\operatorname{Ker}\theta = \{\pm 1\}$. Thus $[F^\star\colon F^{\times 2}] = 2$, and the two cosets are the squares and the nonsquares in $F^\star$.

If $b \in F^\star$ is a nonsquare let $K \supseteq F$ be a splitting field for $x^2 - b$, so $[K\colon F] = 2$ and $|K| = q^2$. Thus $K = \{a + c\sqrt{b}\colon a,\, c \in F\}$, and the norm $N\colon K \to F$ is given by $N(a + c\sqrt{b}) = a^2 - bc^2$. But also

$$N(a + c\sqrt{b}) = (a + c\sqrt{b})^{q+1},$$

since the map $d\colon \mapsto d^q$ generates the Galois group $\mathcal{G}(K\colon F)$. Note that $N\colon K^\star \to F^\star$ is a homomorphism, and that $\operatorname{Ker} N = \{\alpha \in K^\star\colon \alpha^{q+1} = 1\}$, the unique subgroup of order $q + 1$ in $K^\star$. As a consequence the image of N has order $(q^2 - 1)/(q + 1) = q - 1$, and N maps $K^\star$ *onto* $F^\star$.

Note in general that if B is a symmetric form and $0 \neq a \in F$ then B^a, defined by $B^a(u,v) = B(u,v)/a$, is also a symmetric form, and that $Q^a(v) = Q(v)/a$ for all V. The transformation $Q \mapsto Q^a$ is called *scaling* the form by a. Note that B is universal if and only if B^a is universal for all $a \in F^\star$.

Proposition 4.8. *If F is finite, and B is a nondegenerate symmetric bilinear form on a vector space V of dimension $n \geq 2$ over F, then B is universal.*

Proof. It will suffice to prove this for $n = 2$, and by Proposition 4.7 we may assume that B is anisotropic. By Theorem 4.2 we may assume an orthogonal basis $\{v_1, v_2\}$ for which $\widehat{B} = \begin{bmatrix} a & 0 \\ 0 & -b \end{bmatrix}$, $a \neq 0$ and $b \neq 0$, and by scaling we may assume that $a = 1$, so $\widehat{B} = \begin{bmatrix} 1 & 0 \\ 0 & -b \end{bmatrix}$. If $0 \neq v = av_1 + cv_2 \in V$ then $Q(v) = a^2 - bc^2 \neq 0$ since B is anisotropic. Thus b is a nonsquare in F. Setting $K = F(\sqrt{b})$ we have $Q(v) = N_{K \to F}(a + c\sqrt{b})$. It was observed above that N maps $K^\star$ onto $F^\star$, so B is universal. □

Theorem 4.9. *If F is finite* (char $F \neq 2$), *and B is a nondegenerate symmetric bilinear form on an F-space V of dimension $n \geq 2$, then there is a basis for V relative to which $\widehat{B} = \operatorname{diag}(1, \dots, 1, d)$, $0 \neq d \in F$.*

Proof. By **Note II** following Theorem 4.2 and by Proposition 4.8 we may choose $v_1 \in V$ with $Q(v_1) = 1$. We may then continue choosing elements v_i in an orthogonal basis, with $Q(v_i) = 1$, for the same reasons, so long as $\langle v_1, \dots, v_{i-1}\rangle^\perp$ has dimension at least 2, i.e. for $i = 2, \dots, n-1$. There is no longer a guarantee of universality when v_n is to be chosen, so simply choose $v_n \in \langle v_1, \dots, v_{n-1}\rangle^\perp$ with $Q(v_n) = d \neq 0$. □

Note that in the theorem above $\operatorname{discr}(B) = d \cdot F^{\times 2}$.

Corollary 4.10. *Two nondegenerate symmetric bilinear forms B_1 and B_2 on vector spaces V_1 and V_2 over a finite field F* (char $F \neq 2$) *are equivalent if and only if* $\dim V_1 = \dim V_2$ *and* $\operatorname{discr}(B_1) = \operatorname{discr}(B_2)$. *Thus up to equivalence there are just two nondegenerate forms.*

We return to the general discussion of symmetric forms.

A vector space V having a nondegenerate symmetric bilinear form is called a *quadratic space.*

Exercise

Suppose that V is a quadratic space over F and that K is an extension field of F. Set $V^K = K \otimes_F V$, a vector space over K. Show that the bilinear form B on V can be extended to a bilinear form B^K on V^K via $B^K(a \otimes u, b \otimes v) = abB(u, v)$ for $a, b \in K$, $u, v \in V$, and show that V^K is also a quadratic space.

As for symplectic spaces a 2-dimensional subspace H of a quadratic space V is called a *hyperbolic plane* if it has a basis $\{u, v\}$ with $Q(u) = Q(v) = 0$, $B(u, v) = 1$. The basis $\{u, v\}$ is called a *hyperbolic pair* for H. Note that then $\widehat{B}_H = \begin{bmatrix} 0 & 1 \\ 1 & 0 \end{bmatrix}$, and that $\operatorname{discr}(B_H) = -1 \cdot F^{\times 2}$.

Recall that a space V is *isotropic* if $Q(v) = 0$ for some $v \neq 0$ in V. If $Q(v) = 0$ for *all* $v \in V$ then V is called *totally isotropic.* For example, the (one-dimensional) subspace spanned by an isotropic vector is clearly totally isotropic.

Proposition 4.11. *If V is a quadratic space of dimension 2 then the following are equivalent:*

1. *V is a hyperbolic plane,*
2. $\operatorname{discr}(B) = -1 \cdot F^{\times 2}$,
3. *V is isotropic.*

Proof. $1 \Rightarrow 2$: Clear.

$2 \Rightarrow 3$: Since $\operatorname{discr}(B) = -1 \cdot F^{\times 2}$ there is a basis $\{u, v\}$ for V relative to which $\widehat{B} = \begin{bmatrix} b_1 & 0 \\ 0 & b_2 \end{bmatrix}$, with $b_1 b_2 = -c^2$, $c \in F^\star$. Set $w = cu + b_1 v \neq 0$. Then $Q(w) = b_1 c^2 + b_2 b_1^2 = b_1 c^2 - b_1 c^2 = 0$.

$3 \Rightarrow 1$: V is not totally isotropic since it is nondegenerate. Choose nonzero $u \in V$ with $Q(u) = 0$. Since $u \notin V^\perp = 0$ we may choose $v \in V$ with $B(u, v) = a \neq 0$. Note that $\{u, v\}$ is a basis for V since $B(u, u) = 0$. Replace u by $u_1 = a^{-1} u$, so $B(u_1, v) = 1$. For any $b \in F$ we have

$$Q(bu_1 + v) = B(bu_1 + v, bu_1 + v) = 2b + Q(v).$$

Set $v_1 = -\frac{1}{2} Q(v) u_1 + v$; then $Q(v_1) = -Q(v) + Q(v) = 0$, and $\{u_1, v_1\}$ is a hyperbolic pair. □

Exercises

1. Show that a quadratic hyperbolic plane has exactly two totally isotropic subspaces of dimension one.
2. If V is a quadratic space over a finite field F and $\dim V \geq 3$ show that V is isotropic.

The final proposition in this chapter is rather technical – it will prove useful in the sequel.

Proposition 4.12. *Suppose that V is a quadratic space and that U is a subspace with* $\operatorname{rad} U \neq 0$. *Let U' be any complementary subspace for* $\operatorname{rad} U$ *in U, i.e.* $U = \operatorname{rad} U \mathbin{\bigcirc\!\!\!\perp} U'$. *If $\{u_1, \ldots, u_k\}$ is a basis for* $\operatorname{rad} U$ *then there is a subspace W, with basis $\{v_1, \ldots, v_k\}$, such that $U \cap W = 0$, $U \oplus W$ is nondegenerate, (u_i, v_i) is a hyperbolic pair for $H_i = \langle u_i, v_i \rangle$, $1 \leq i \leq k$, and*

$$U \oplus W = U' \mathbin{\bigcirc\!\!\!\perp} H_1 \mathbin{\bigcirc\!\!\!\perp} H_2 \mathbin{\bigcirc\!\!\!\perp} \cdots \mathbin{\bigcirc\!\!\!\perp} H_k .$$

Proof. Use induction on $k = \dim \operatorname{rad} U$. Note that $\operatorname{rad} U = \operatorname{rad} U \oplus \operatorname{rad} U'$, so $\operatorname{rad} U' = 0$ and U' is a nondegenerate subspace. Define $f \in U^*$ via

$$f(u_1) = 1,\ f(u_i) = 0 \text{ for } 2 \leq i \leq k, \text{ and } f|_{U'} \equiv 0.$$

By Corollary 2.3 there exists $v_1 \in V$ with $f = R_{v_1}|_U$, so $B(u_1, v_1) = 1$, $B(u_i, v_1) = 0$ for $2 \leq i \leq k$, and $v_1 \perp U'$. The same is true if we replace v_1 by $v_1 + au_1$ for any $a \in F$, so we may assume that $Q(v_1) = 0$, hence that (u_1, v_1) is a hyperbolic pair. Set $H_1 = \langle u_1, v_1 \rangle$, then $V = H_1 \perp H_1^{\perp}$. If $k = 1$ we may set $W = \langle v_1 \rangle$ and then $U \oplus W = U' \perp H_1$ as needed.

Suppose then that $k > 1$. Set $V_1 = H_1^{\perp}$ and set

$$U_1 = U' \perp \langle u_2, \dots, u_k \rangle \subseteq V_1.$$

Clearly $\operatorname{rad} U_1 = \langle u_2, \dots, u_k \rangle$. By induction there are $v_2, \dots, v_k \in V$ such that $v_i \perp U'$, (u_i, v_i) is a hyperbolic pair for the hyperbolic plane $H_i = \langle u_i, v_i \rangle$, $2 \leq i \leq k$, $U_1 \cap \langle v_2, \dots, v_k \rangle = 0$, and

$$U_1 \oplus \langle v_2, \dots, v_k \rangle = U' \perp H_2 \perp \cdots \perp H_k.$$

Set $W = \langle v_1, \dots, v_k \rangle$. Then $U \cap W = 0$ and

$$U \oplus W = U' + \langle u_1, \dots, u_k \rangle + \langle v_1, \dots, v_k \rangle = U' \perp H_1 \perp \cdots \perp H_k,$$

and the proof is complete. □

Chapter 5

Orthogonal Geometry (char $F \neq 2$)

Throughout this chapter we assume that V is a quadratic space of dimension $n \geq 2$ over a field F, with char $F \neq 2$. The isometries of V are called *orthogonal (linear) transformations*; they comprise the *orthogonal group* $O(V)$. Thus

$$O(V) = \{\tau \in \mathrm{GL}(V) \colon B(\tau u, \tau v) = B(u, v), \text{ all } u,\, v \in V\} \leq \mathrm{GL}(V).$$

Equivalently $\tau \in O(V)$ if and only if $Q(\tau v) = Q(v)$, all $v \in V$.

If $\{v_1, \ldots, v_n\}$ is a basis for V and $\tau \in \mathrm{GL}(V)$ then, just as for $\mathrm{Sp}(V)$, we see that $\tau \in \mathrm{O}(V)$ if and only if $T^t \widehat{B} T = \widehat{B}$, where T is the matrix representing τ relative to the basis. As a result

$$(\det T)^2 \cdot \det \widehat{B} = \det \widehat{B},$$

so $\det T = \pm 1$ for any $\tau \in \mathrm{O}(V)$. If $\det \tau = 1$ call τ a *rotation*, or a *proper* orthogonal transformation. If $\det \tau = -1$ call τ a *reversion*, or (more commonly) an *improper* orthogonal transformation.

The rotations in $\mathrm{O}(V)$ form a subgroup which will be denoted $\mathrm{SO}(V)$, called the *special orthogonal group* – it is often denoted by $\mathrm{O}^+(V)$ or $\mathrm{O}_+(V)$ in the literature.

We shall see shortly that improper orthogonal transformations exist in $\mathrm{O}(V)$. Thus the determinant function is a homomorphism from $\mathrm{O}(V)$ *onto* $\{\pm 1\} \leq F^\star$ with kernel $\mathrm{SO}(V)$, so $[\mathrm{O}(V) \colon \mathrm{SO}(V)] = 2$.

Let $u \in V$ be any nonzero anisotropic vector (i.e. $Q(u) \neq 0$), and define a linear transformation σ_u via

$$\sigma_u(v) = v - 2\frac{B(v,u)}{Q(u)} \cdot u$$

for all $v \in V$. Then

$$\begin{aligned} B(\sigma_u v, \sigma_u w) &= B(v - 2\frac{B(v,u)}{Q(u)}u, w - 2\frac{B(w,u)}{Q(u)}u) \\ &= B(v,w) - 4\frac{B(v,u)B(w,u)}{Q(u)} + 4\frac{B(v,u)B(w,u)Q(u)}{Q(u)^2} \\ &= B(v,w), \end{aligned}$$

so $\sigma_u \in \mathrm{O}(V)$.

Note that $\sigma(u) = -u$, and if $v \perp u$ then $\sigma_u v = v$. We call σ_u the *(orthogonal) reflection along* u or *through* the hyperplane $u^\perp$. Often in the literature σ_u is called a *symmetry*, e.g. by Artin in [**3**].

Since $Q(u) \neq 0$ we have $V = \langle u \rangle \oplus \langle u \rangle^\perp$ by Proposition 2.9. If we set $u_1 = u$ and choose any basis $\{u_2, \dots, u_n\}$ for $\langle u \rangle^\perp$, then relative to the basis $\{u_1, u_2, \dots, u_n\}$ for V the matrix representing σ_u is $\begin{bmatrix} -1 & 0 \\ 0 & I_{n-1} \end{bmatrix}$, so σ_u is improper. Clearly σ_u is of order 2.

Exercises

1. If $u \in V$ is anisotropic and $\tau \in \mathrm{O}(V)$ show that $\tau\sigma_u\tau^{-1} = \sigma_{\tau u}$.
2. If H is a hyperbolic plane and $\tau \in \mathrm{O}(H)$ fixes an isotropic vector in H show that $\tau = 1_H$.

Theorem 5.1 (Witt's Cancellation Theorem). *Suppose that U_1 and U_2 are nondegenerate subspaces of a quadratic space V and that $\sigma\colon U_1 \to U_2$ is an isometry. Then $U_1^\perp$ and $U_2^\perp$ are also isometric.*

Proof. Induction on $\dim U_1$. To begin suppose that $U_1 = \langle u_1 \rangle$, $U_2 = \langle u_2 \rangle$, and hence $Q(u_i) \neq 0$ by nondegeneracy. We may assume that $\sigma(u_1) = \sigma(u_2)$, and so $Q(u_1) = Q(u_2)$. Then

$$Q(u_1 \pm u_2) = 2Q(u_1) \pm 2B(u_1, u_2),$$

so if $Q(u_1 + u_2) = Q(u_1 - u_2) = 0$ then $Q(u_1)$ is equal to both of $B(u_1, u_2)$ and $-B(u_1, u_2)$, contradicting $Q(u_1) \neq 0$. Thus one, at least, of $Q(u_1 + u_2)$ and $Q(u_1 - u_2)$ is nonzero.

Say that $Q(u_1 + u_2) \neq 0$. Note then that

$$B(u_1 + u_2, u_1 - u_2) = Q(u_1) - Q(u_2) = 0,$$

so $u_1 - u_2 \perp u_1 + u_2$, and $\sigma_{u_1+u_2}(u_1 - u_2) = u_1 - u_2$. Thus

$$\begin{aligned} \sigma_{u_1+u_2}(u_1) &= \sigma_{u_1+u_2}(\frac{1}{2}(u_1+u_2) + \frac{1}{2}(u_1-u_2)) \\ &= -\frac{1}{2}(u_1+u_2) + \frac{1}{2}(u_1-u_2) \\ &= -u_2. \end{aligned}$$

Thus $\sigma_{u_1+u_2}(\langle u_1 \rangle) = \langle u_2 \rangle$, and $\sigma_{u_1+u_2}(\langle u_1 \rangle^{\perp}) = \langle u_2 \rangle^{\perp}$, since

$$\sigma_{u_1+u_2} \in \mathrm{O}(V).$$

The proof is similar, using $\sigma_{u_1-u_2}$ instead of $\sigma_{u_1+u_2}$, if $Q(u_1 - u_2) \neq 0$.

Suppose next that $\dim U_1 > 1$ and assume the result for subspaces of lower dimensions. Choose u_1 anisotropic in U_1 and let W_1 be the orthogonal complement of u_1 in U_1. Then W_1 is nondegenerate and $U_1 = \langle u_1 \rangle \operatorname{\bigcirc\!\!\!\!\perp} W_1$. Set $u_2 = \sigma u_1$ and $W_2 = \sigma W_2$, so $U_2 = \langle u_2 \rangle \operatorname{\bigcirc\!\!\!\!\perp} W_2$. Then

$$V = \langle u_1 \rangle \operatorname{\bigcirc\!\!\!\!\perp} W_1 \operatorname{\bigcirc\!\!\!\!\perp} U_1^{\perp} = \langle u_2 \rangle \operatorname{\bigcirc\!\!\!\!\perp} W_2 \operatorname{\bigcirc\!\!\!\!\perp} U_2^{\perp}.$$

By the 1-dimensional case above there is an isometry η from $W_1 \operatorname{\bigcirc\!\!\!\!\perp} U_1^{\perp}$ to $W_2 \operatorname{\bigcirc\!\!\!\!\perp} U_2^{\perp}$. Thus $W_2 \operatorname{\bigcirc\!\!\!\!\perp} U_2^{\perp} = \eta W_1 \operatorname{\bigcirc\!\!\!\!\perp} \eta(U_1^{\perp})$, and $\eta\sigma^{-1}$ is an isometry from W_2 to ηW_1. Also $U_2^{\perp}$ and $\eta(U_1^{\perp})$ are the orthogonal complements in $V_1 = W_2 \operatorname{\bigcirc\!\!\!\!\perp} U_2^{\perp}$ of W_2 and ηW_1, respectively, so by induction $U_2^{\perp}$ and $\eta(U_1^{\perp})$ are isometric, hence $U_1^{\perp}$ and $U_2^{\perp}$ are isometric. □

Theorem 5.2 (Witt's Extension Theorem). *If U_1 and U_2 are subspaces of a quadratic space V and $\sigma\colon U_1 \to U_2$ is an isometry, then there exists $\tau \in \mathrm{O}(V)$ with $\tau|_{U_1} = \sigma$.*

Proof. Suppose first that U_1 and U_2 are nondegenerate. By Theorem 5.1 there is an isometry $\eta\colon U_1^{\perp} \to U_2^{\perp}$. Since $V = U_1 \oplus U_1^{\perp} = U_2 \oplus U_2^{\perp}$ it is clear that $\tau = \sigma \oplus \eta \in \mathrm{O}(V)$, and that $\tau|_{U_1} = \sigma$.

Suppose then that $\operatorname{rad} U_1 \neq 0$, and write $U_1 = \operatorname{rad} U_1 \operatorname{\bigcirc\!\!\!\!\perp} U_1'$ for some subspace U_1'. By Proposition 4.12 there is a subspace W_1 such that $U_1 \oplus W_1$ is nondegenerate and $U_1 \oplus W_1 = U_1' \operatorname{\bigcirc\!\!\!\!\perp} H_1 \operatorname{\bigcirc\!\!\!\!\perp} \cdots \operatorname{\bigcirc\!\!\!\!\perp} H_k$, with each H_i a hyperbolic plane with hyperbolic pair (u_i, v_i). Similarly we have W_2 with $U_2 \oplus W_2$ nondegenerate and $U_2 \oplus W_2 = U_2' \operatorname{\bigcirc\!\!\!\!\perp} H_1' \operatorname{\bigcirc\!\!\!\!\perp} \cdots \operatorname{\bigcirc\!\!\!\!\perp} H_k'$, with $U_2' = \sigma U_1'$, $H_i' = \langle \sigma u_i, v_i' \rangle$, $(\sigma u_i, v_i')$ a hyperbolic pair. Extend σ to $\sigma' : U_1 \oplus W_1 \to U_2 \oplus W_2$ via $\sigma'(v_i) = v_i'$; clearly σ' is an isometry. Since $U_1 \oplus W_1$ is nondegenerate there exists $\tau \in \mathrm{O}(V)$ with $\tau|_{U_1 \oplus W_1} = \sigma'$ by the first part of the proof, and hence $\tau|_{U_1} = \sigma$. □

Corollary 5.3. *Any two maximal totally isotropic subspaces of V have the same dimension, and every totally isotropic subspace is contained in one of maximal dimension.*

Proof. Let U be a totally isotropic subspace of maximal dimension, say m. If W is any totally isotropic subspace then there is an isometry σ from W to a subspace of U. By the theorem there exists $\tau \in \mathrm{O}(V)$ with $\tau|_U = \sigma$. But then $\tau^{-1}U$ is a totally isotropic subspace of dimension m, maximal, and $W \subseteq \tau^{-1}U$. □

The dimension m of a maximal totally isotropic subspace of a quadratic space V is called the *Witt index* of V; it is denoted variously as $m(V)$, or $m(B)$, or $m(Q)$, or sometimes in the literature by $\mathrm{ind}(V)$, etc.

A subspace H of a quadratic space V is called *hyperbolic* if H is an orthogonal direct sum of hyperbolic planes.

Theorem 5.4. *If V is a quadratic space with Witt index m then V has a hyperbolic subspace H of dimension $2m$ and an anisotropic subspace X with $V = H \oplus X$. If $V = H' \oplus Y$ is any orthogonal splitting with H' hyperbolic and Y anisotropic then* $\dim H' = 2m$*, and Y and X are isometric.*

Proof. Choose a totally isotropic subspace U with $\dim U = m$. By Proposition 4.12 V has a subspace W such that $U \cap W = 0$, $H = U \oplus W$ is hyperbolic, and $\dim H = 2m$. Then $V = H \oplus H^{\perp}$; set $X = H^{\perp}$. Clearly $X = H^{\perp}$ is anisotropic, since $U \perp X$ and $\dim U = m$ is maximal.

Suppose that $H' = H_1' \oplus H_2' \cdots \oplus H_k'$, each H_i' a hyperbolic plane with hyperbolic pair (u_i, v_i). Then $\langle u_1, \ldots, u_k \rangle$ is totally isotropic, so $k \leq m$. Also $H_1 \oplus \cdots \oplus H_k$ and $H_1' \oplus \cdots \oplus H_k'$ are isometric, so by Witt's Extension Theorem (5.2) there exists $\tau \in \mathrm{O}(V)$ carrying $H_1 \oplus \cdots H_k$ to $H_1' \oplus \cdots H_k'$. Thus τ^{-1} carries $(\sum_1^k H_i)^{\perp} = H_{k+1} \oplus \cdots \oplus H_m \oplus X$ to $(\sum_1^k H_i')^{\perp} = Y$, and hence $k = m$, since Y is anisotropic. It follows that X and Y are isometric by the Witt Cancellation Theorem (5.1). □

An anisotropic subspace such as X or Y in Theorem 5.4 is called an *anisotropic kernel* for V; it is uniquely determined up to isometry.

Corollary 5.5. *Quadratic spaces V_1 and V_2 of the same dimension over F are equivalent if and only if their anisotropic kernels are equivalent.*

Corollary 5.6. *If V has dimension n and Witt index m then $m \leq n/2$.*

Proposition 5.7. *Suppose that V is a quadratic space, W is a hyperplane in V, $\tau \in \mathrm{SO}(V)$, and $\tau|_W = 1_W$. Then $\tau = 1_V$.*

Proof. If W is nondegenerate then $V = W \oplus W^{\perp}$, and $\dim W^{\perp} = 1$, so τ can be represented by a matrix of the form $\begin{bmatrix} I_{n-1} & 0 \\ 0 & a \end{bmatrix}$, $a \in F$. Thus $1 = \det \tau = a$, and $\tau = 1_V$.

If W is degenerate then $W \cap W^{\perp} \neq 0$, so $W^{\perp} \subseteq W$ and $\operatorname{rad} W = W^{\perp} = \langle u \rangle$, say. Write $W = U \textcircled{\perp} \langle u \rangle$. By Proposition 4.12 there exists $v \in V$ such that $H = \langle u, v \rangle$ is a hyperbolic plane with (u, v) as a hyperbolic pair, and $U \oplus H = V$. But $\tau|_U = 1_U$ and τ fixes the isotropic vector u in H, so $\tau|_H = 1_H$ by Exercise 2 on page 40. Thus $\tau = 1_V$. □

Let us use some of the ideas above to develop, in the exercises below, some interesting group theory, which will not be required in later chapters, although the concepts will be mentioned briefly in Chapter 15.

If $p \in \mathbb{N}$ is a prime then a finite p-group G is called *extra-special* if $G' = Z(G) = Z = \langle z \rangle$ is of order p and G/Z is elementary abelian. For example both nonabelian groups of order p^3 are extra-special.

Assume for the following exercises that p is an odd prime and that G is an extra-special p-group.

Exercises

1. View $V = G/Z$ as an $\mathbb{F}_p$-vector space. If $x, y \in G$ and the commutator $[x, y]$ is z^k, $0 \leq k \leq p-1$ define $B(xZ, yZ) = k \cdot 1 \in \mathbb{F}_p$. Show that B is a nondegenerate symplectic bilinear form on V.
2. Show that $|V| = p^{2m}$ and $|G| = p^{2m+1}$ for some $m \in \mathbb{N}$ (V is an orthogonal direct sum of hyperbolic planes). Show that there are subgroups $G_i \triangleleft G$, $1 \leq i \leq m$, with G_i extra-special of order p^3, $[G_i, G_j] = 1$ and $G_i \cap G_j = Z$ if $i \neq j$, and $\prod_1^m G_i = G$ (we say that G is the *central product* of the G_i's). *Suggestion:* Lift the hyperbolic planes back to G.
3. Show that A is a maximal abelian normal subgroup of G if and only if A/Z is a maximal totally isotropic subspace of V. Conclude in that case that $|A| = p^{m+1}$.

For further development of these ideas, including extension of the results to $p = 2$, see the discussion beginning on page 75 of [**42**].

Chapter 6

Orthogonal Groups (char $F \neq 2$), I

We continue to assume that V is a quadratic space of dimension $n = 2$ or greater over a field F of characteristic different from 2. Further restrictions will be imposed later in the chapter.

Proposition 6.1. *If V is a hyperbolic plane then* $\mathrm{SO}(V) \cong F^\star$, *every improper transformation in* $\mathrm{O}(V)$ *is a reflection, and* $\mathrm{O}(V)$ *is a semidirect product* $\mathrm{SO}(V) \rtimes \langle \sigma_u \rangle$ *for any reflection* $\sigma_u \in \mathrm{O}(V)$. *If F is finite then* $\mathrm{O}(V)$ *is dihedral of order* $2(q-1)$, *where* $|F| = q$.

Proof. If (u, v) is a hyperbolic pair for V then $\langle u \rangle$ and $\langle v \rangle$ are the two unique isotropic lines in V (see the exercise on page 37). Take $\tau \in \mathrm{O}(V)$. Then either *(i)* $\tau(\langle u \rangle) = \langle u \rangle$ and $\tau(\langle v \rangle) = \langle v \rangle$, or *(ii)* $\tau(\langle u \rangle) = \langle v \rangle$ and $\tau(\langle v \rangle) = \langle u \rangle$.

In case *(i)* say $\tau u = au$ and $\tau v = bv$. Then

$$1 = B(u, v) = B(\tau u, \tau v) = abB(u, v) = ab,$$

so $b = a^{-1}$. Thus τ is represented by $\begin{bmatrix} a & 0 \\ 0 & a^{-1} \end{bmatrix}$, and τ is a rotation.

In case *(ii)* say $\tau u = av$ and $\tau v = bu$. Again $b = a^{-1}$, but this time τ is represented by $\begin{bmatrix} 0 & a^{-1} \\ a & 0 \end{bmatrix}$, and τ is improper.

Conversely, transformations defined by those matrices are in $\mathrm{O}(V)$, and are rotations or are improper, respectively. The map

$$\begin{bmatrix} a & 0 \\ 0 & a^{-1} \end{bmatrix} \mapsto a$$

provides an isomorphism from $\mathrm{SO}(V)$ to $F^\star$.

If τ is improper, with $\tau u = av$ and $\tau v = a^{-1}u$, set $w = u - av$ and $w' = u + av$, and note that $\langle w \rangle^\perp = \langle w' \rangle$. Since $\tau w = -w$ and $\tau w' = w'$ it is clear that τ is the reflection σ_w.

If σ_u is a reflection, and σ_w is any other reflection, then $\sigma_w \sigma_u \in \mathrm{SO}(V)$, so $\sigma_w \in \mathrm{SO}(V)\sigma_u$. Thus $\mathrm{SO}(V)$ is of index 2 in $\mathrm{O}(V)$, hence is normal, and it is clear that $\mathrm{O}(V) = \mathrm{SO}(V) \rtimes \langle \sigma_u \rangle$.

If F is finite then $F^\star$ is cyclic, so $\mathrm{SO}(V)$ is cyclic. It is easy to check, using the representing matrices above, that each reflection conjugates every rotation to its inverse, and so $\mathrm{O}(V)$ is dihedral of order $2(q-1)$. □

Let us develop the analogous results of Proposition 6.1 for anisotropic planes in the following exercises. Assume, then, that V is an anisotropic quadratic space of dimension 2 over F. Observe that without changing $\mathrm{O}(V)$ we may assume (by scaling) that $\widehat{B} = \begin{bmatrix} 1 & 0 \\ 0 & -d \end{bmatrix}$ relative to a basis $\{v_1, v_2\}$, where $d \in F^\star \setminus F^{\times 2}$. Set $K = F(\sqrt{d})$.

Exercises

1. Show that K is a 2-dimensional quadratic space over F with quadratic form the norm map $N_{K/F}$. Define $\eta\colon V \to K$ via

$$\eta(a_1 v_1 + a_2 v_2) = a_1 + a_2\sqrt{d}$$

and show that η is an isometry.

2. If $c \in K$ with $N(c) = 1$ define $\rho_c\colon K \to K$ via $\rho_c(a) = ca$, all $a \in K$. Show that $\rho_c \in \mathrm{SO}(K)$, and conversely if $\sigma \in \mathrm{SO}(K)$ then $\sigma = \rho_c$, where $c = \sigma(1)$. Conclude that $\mathrm{SO}(V) \cong \operatorname{Ker} N_{K^\star/F^\star}$, so $\mathrm{SO}(V)$ is abelian. Show in fact that $\mathrm{SO}(V) \cong K^\star/F^\star$ (see Hilbert's Satz 90, e.g. in **[28]**, page 118, or **[44]**, page 286).

3. Define $\tau\colon K \to K$ via $\tau(a_1 + a_2\sqrt{d}) = a_1 - a_2\sqrt{d}$. Show that $\tau \in \mathrm{O}(K)$, $\tau^2 = 1$, $\tau\rho_c\tau^{-1} = \rho_c^{-1}$ for all $c \in K$, and that $\mathrm{O}(K)$ is the semidirect product of $\mathrm{SO}(K)$ and $\langle \tau \rangle$. If $|F| = q$, finite, conclude that $\mathrm{O}(2, q)$ is dihedral of order $2(q+1)$.

Proposition 6.2. *If* $\tau \in \mathrm{O}(V)$ *then* $\operatorname{Ker}(\tau - 1_V) = \operatorname{Im}(\tau - 1_V)^\perp$.

Proof. If $v \in \operatorname{Ker}(\tau - 1)$ and $w \in V$ then

$$B(v, (\tau - 1)w) = B(v, \tau w) - B(v, w) = B(\tau^{-1}v, w) - B(v, w) = 0,$$

since $\tau v = v$ and hence $v = \tau^{-1}v$. Thus $\mathrm{Ker}(\tau - 1) \subseteq \mathrm{Im}(\tau - 1)^{\perp}$. On the other hand take $v \in \mathrm{Im}(\tau - 1)^{\perp}$ and any $w \in V$. Then

$$\begin{aligned} B((\tau - 1)v, \tau w) &= B(\tau v, \tau w) - B(v, \tau w) \\ &= B(v, w) - B(v, \tau w) \\ &= -B(v, (\tau - 1)w) = 0, \end{aligned}$$

and so $(\tau - 1)v \in \mathrm{rad}\, V = 0$, i.e. $v \in \mathrm{Ker}(\tau - 1)$. □

Corollary 6.3. *If $\tau \in \mathrm{O}(V)$ then $(\tau - 1)^2 = 0$ if and only if $\mathrm{Im}(\tau - 1)$ is totally isotropic.*

Proof. Note that $\mathrm{Im}(\tau - 1)$ is totally isotropic if and only if $\mathrm{Im}(\tau - 1) \subseteq \mathrm{Im}(\tau - 1)^{\perp}$, which by the theorem is if and only if $\mathrm{Im}(\tau - 1) \subseteq \mathrm{Ker}(\tau - 1)$ if and only if $(\tau - 1)^2 = 0$. □

Proposition 6.4. *If $\tau \in \mathrm{O}(V)$ and $0 \neq v \in V$ then $(\tau - 1)v$ is isotropic if and only if $(\tau - 1)v \perp v$.*

Proof. Note that

$$B(\tau v - v, \tau v - v) = 2[B(v, v) - B(\tau v, v)] = -2B(\tau v - v, v).$$

The result follows. □

Proposition 6.5. *Suppose that $\tau \in \mathrm{O}(V)$, with $(\tau - 1)^2 \neq 0$. Then there is a nonzero anisotropic vector $v \in V$ such that $z = \tau v - v$ is anisotropic. If $z \neq 0$ then $\sigma_z \tau v = v$.*

Proof. If not then for every nonzero anisotropic vector $v \in V$ we must have $\tau v - v$ isotropic (in particular nonzero), and hence $\tau v - v \perp v$. It follows that the subspace $\langle v\,, \tau v - v\rangle$ is degenerate, since $\tau v - v$ is in its radical. Since V is nondegenerate $\dim V$ must be at least 3. We show next that $\tau w - w \perp w$ for all $w \in V$. This has been observed already if w is anisotropic, so suppose w is isotropic. Choose $x \neq 0$, anisotropic, in $\langle w\rangle^{\perp}$ (e.g. apply Proposition 4.12), and set $u = w + ax$, $a \in F^{\star}$. Then $Q(u) = a^2 Q(x) \neq 0$, so u is anisotropic and nonzero. Thus $\tau u \perp u$, which says that

$$\begin{aligned} 0 &= B(\tau w + a\tau x - w - ax, w + ax) \\ &= B(\tau w - w, w) + a[B(\tau x - x, w) + B(\tau w - w, x)]. \end{aligned}$$

Since $a \in F^{\star}$ is arbitrary the final expression enclosed in square brackets must be 0, and therefore $B(\tau w - w, w) = 0$. By Proposition 6.4 we may conclude that $\mathrm{Im}(\tau - 1)$ is totally isotropic, and then conclude that $(\tau - 1)^2 =$

0 by Corollary 6.3, a contradiction. Thus in fact there exists $v \neq 0$ in V, v anisotropic, such that $z = \tau v - v$ is also anisotropic. If $z \neq 0$ then

$$\begin{aligned} \sigma_z \tau v &= \sigma_z(z+v) \\ &= -z + v - 2\frac{B(v,z)}{Q(z)} z \\ &= v - \left[\frac{B(z,z) + B(2v,z)}{Q(z)}\right] z \\ &= v - \frac{B(\tau v + v, z)}{Q(z)} z \\ &= v - \frac{B(\tau v + v, \tau v - v)}{Q(z)} \cdot z \;=\; v. \end{aligned}$$

□

Theorem 6.6. *(E. Cartan–Dieudonné) If V is a quadratic space of dimension n then every $\tau \in \mathrm{O}(V)$ is a product of n or fewer reflections.*

Proof. We use induction on n. The result is immediate when $n = 1$, since $\mathrm{O}(V) = \{\pm 1_V\}$ in that case, and -1_V is a reflection (we view 1_V as a product of 0 reflections). Assume the result for dimensions less than n, and suppose that $\tau \in \mathrm{O}(V)$ is a counterexample at dimension n. We show first that $(\tau - 1)^2 = 0$. If not, choose $v \neq 0$, anisotropic, such that $z = \tau v - v$ is also anisotropic, as in Proposition 6.5. If $z = 0$ then $\tau v = v$, so $\tau|_{\langle v \rangle^\perp} \in \mathrm{O}(\langle v \rangle^\perp)$, and by induction $\tau|_{\langle v \rangle^\perp}$ is a product of $n-1$ or fewer reflections in $\mathrm{O}(\langle v \rangle^\perp)$, each of which extends to a reflection in $\mathrm{O}(V)$ fixing v, and τ is a product of $n-1$ or fewer reflections.

If $z \neq 0$ then v is fixed by $\sigma_z \tau$, by Proposition 6.5. As above $\sigma_z \tau$ is a product of $n-1$ or fewer reflections in $\mathrm{O}(v^\perp)$, each of which extends to V, and hence τ is a product of n or fewer reflections, a contradiction. Thus in fact $(\tau - 1)^2 = 0$, and hence $\mathrm{Im}(\tau - 1) \subseteq \mathrm{Ker}(\tau - 1)$. By the Corollary to Proposition 6.2 $\mathrm{Im}(\tau - 1)$ is totally isotropic. In fact, $\mathrm{Ker}(\tau - 1)$ must be totally isotropic, for otherwise τ is a product of $n-1$ or fewer reflections, exactly as in the case $z = 0$ above. Thus we have

$$\mathrm{Ker}(\tau - 1) \subseteq \mathrm{Ker}(\tau - 1)^\perp = \mathrm{Im}(\tau - 1),$$

by Proposition 6.2, and so

$$\mathrm{Im}(\tau - 1) = \mathrm{Ker}(\tau - 1) = \mathrm{Ker}(\tau - 1)^\perp.$$

Consequently

$$\begin{aligned} n &= \dim \mathrm{Ker}(\tau - 1) + \dim\, \mathrm{Ker}(\tau - 1)^\perp \\ &= \dim\, \mathrm{Fix}(\tau) + \dim\, \mathrm{Fix}(\tau) = 2 \cdot \dim\, \mathrm{Fix}(\tau), \end{aligned}$$

and n is even.

Note that $\tau|_{\mathrm{Fix}(\tau)} = 1_{\mathrm{Fix}(\tau)}$, and also that τ induces the identity map on $V/\mathrm{Fix}(\tau)$. (The reason: if we write $W = \mathrm{Fix}(\tau)$, then $\tau(v+W) = \tau v + W = v + \tau v - v + W = v + W$). It follows that $\tau \in \mathrm{SO}(V)$. Let σ be some reflection in $\mathrm{O}(V)$, so $\sigma\tau \in \mathrm{O}(V) \setminus \mathrm{SO}(V)$. If $\sigma\tau$ were also a counterexample to the theorem then the arguments above would lead to the conclusion that $\sigma\tau \in \mathrm{SO}(V)$, a contradiction. Thus $\sigma\tau$ is a product of n or fewer reflections, and hence τ is a product of $n+1$ or fewer reflections.

But n is even and $\tau \in \mathrm{SO}(V)$, so τ is in fact a product of n or fewer reflections, a final contradiction. □

Corollary 6.7. *Suppose $\tau \in \mathrm{O}(V)$ is a product of $r \leq n$ reflections. Then $\dim(\mathrm{Fix}\,\tau) \geq n - r$.*

Proof. If $\tau = \sigma_1 \cdots \sigma_r$, each σ_i a reflection, set $W_i = \mathrm{Fix}\,\sigma_i$, the hyperplane fixed by σ_i. Then $\mathrm{Fix}\,\tau \supseteq \bigcap_1^r W_i$, and clearly $\dim \bigcap_1^r W_i \geq n - r$. □

Corollary 6.8. *If $\mathrm{Fix}\,\tau = 0$ then τ is not a product of fewer than n reflections.*

Corollary 6.9. *If $\dim V = 2$ then every improper transformation $\tau \in \mathrm{O}(V)$ is a reflection.*

Corollary 6.10. *If $\dim V = 3$ and $1 \neq \tau \in \mathrm{SO}(V)$, so τ is a rotation, then $\dim\,\mathrm{Fix}(\tau) = 1$, i.e. τ has an "axis of rotation".*

Proof. Write $\tau = \sigma_1\sigma_2$, each σ_i a reflection with fixed hyperplane W_i. Then $\mathrm{Fix}\,\tau \supseteq W_1 \cap W_2$. If $\dim\,\mathrm{Fix}(\tau)$ were 2 or greater, then $\mathrm{Fix}\,\tau$ would be all of V, since $\det\tau = 1$. Thus $\mathrm{Fix}\,\tau = W_1 \cap W_2$, of dimension 1. □

We remark that when $F = \mathbb{R}$ and $\widehat{B} = I_3$ (identity matrix) for an appropriate basis we have ordinary Euclidean 3-space, in which case Corollary 6.10 is commonly known as *Euler's Theorem* (see e.g. [**29**], page 10). Euler's Theorem is often cited, and proved, with rather more complicated proofs than that above, in undergraduate texts on Classical Mechanics.

In general, if $V = U \oplus W$, if σ is a linear transformation on U, and τ is a linear transformation on W, then there is an obvious meaning for $\sigma \oplus \tau$, a linear transformation on V. If V is a quadratic space and $V = U \mathbin{\bigcirc\!\!\!\!\perp} W$ we shall write instead $\sigma \mathbin{\bigcirc\!\!\!\!\perp} \tau$. It is clear that if $\sigma \in \mathrm{O}(U)$ and $\tau \in \mathrm{O}(W)$ then $\sigma \mathbin{\bigcirc\!\!\!\!\perp} \tau \in \mathrm{O}(V)$. Note as an example that if $u \neq 0$ is anisotropic in V, and $U = \langle u \rangle$, then $V = U \mathbin{\bigcirc\!\!\!\!\perp} U^\perp$, and the reflection along u is $-1_U \mathbin{\bigcirc\!\!\!\!\perp} 1_{U^\perp}$.

Proposition 6.11. *Suppose that V is a quadratic space and that $\tau \in \mathrm{O}(V)$ is an involution, i.e. $\tau^2 = 1$ (but $\tau \neq 1$). Then V has a nondegenerate subspace U such that $\tau = -1_U \mathbin{\bigcirc\!\!\!\!\perp} 1_{U^\perp}$.*

Proof. Set $U = \mathrm{Im}(\tau - 1)$, then $U^\perp = \mathrm{Ker}(\tau - 1) = \mathrm{Fix}(\tau)$ by Proposition 6.2. If $v \in U \cap U^\perp$ then $v = (\tau - 1)u$ for some $u \in V$ and also $\tau v = v$, so $v = \tau(\tau - 1)u = (1 - \tau)u = -(\tau - 1)u = -v$, and so $v = 0$. Thus U is nondegenerate and $v = U \bigoplus U^\perp$. But $\tau|_U = -1_U$ (as was just shown above), and $\tau|_{U^\perp} = 1_{U^\perp}$, so $\tau = -1_U \bigoplus 1_{U^\perp}$. □

If U (in Proposition 6.11) has dimension m we say that τ is an *involution of type* m. Thus there are n types of involutions in $\mathrm{O}(V)$, beginning with reflections, which are of type 1, through -1_V, the unique involution of type n. An involution of type 2, say $\tau = -1_U \bigoplus 1_{U^\perp}$, where U is a nondegenerate plane in V, will be called a *half-turn* (or a 180° *rotation*), with "axis" $U^\perp$ of dimension $n - 2$. The terminology is at the very least accurately descriptive for 2- and 3-dimensional real Euclidean spaces.

Proposition 6.12. *If u and v are nonzero orthogonal anisotropic vectors in V, and $U = \langle u, v\rangle$, then $\sigma_u\sigma_v = \sigma_v\sigma_u$ is a half-turn with axis $U^\perp$.*

Proof. Clearly u and v are linearly independent, and U is nondegenerate, so $V = U \bigoplus U^\perp$. Observe now that $\sigma_u\sigma_v|_U = -1_U$ (apply $\sigma_u\sigma_v$ to u and to v) and that $\sigma_u\sigma_v|_{U^\perp} = 1_{U^\perp}$, and similarly for $\sigma_v\sigma_u$. □

Proposition 6.13. *If V is a quadratic space of dimension 3 or more, and if $\tau \in \mathrm{SO}(V)$, then τ is a product of $2k$ half-turns, with $2k \leq n$.*

Proof. By the Cartan-Dieudonné Theorem (6.6) τ is a product of $2k \leq n$ reflections (an even number since τ is a rotation). Thus it will suffice to show that each product $\sigma_u\sigma_v$ of two reflections is equal to a product of two half-turns.

If $n = 3$ then $-\sigma_u$ and $-\sigma_v$ are both half-turns, and $\sigma_u\sigma_v = (-\sigma_u)(-\sigma_v)$.

Suppose then that $n \geq 4$. We may assume that $\{u, v\}$ is linearly independent, for otherwise $\sigma_u = \sigma_v$ and $\sigma_u\sigma_v = 1$. Set $W = \langle u, v\rangle$. Then $W^\perp$ cannot be totally isotropic, for otherwise $W^\perp \subseteq W^{\perp\perp} = W$, and

$$\dim W^\perp = n - \dim W \geq 4 - 2 = 2,$$

so $W^\perp = W$, contradicting the fact that $u \in W$, $u \neq 0$, and u is anisotropic. Thus there exists an anisotropic $w \in W^\perp$, $w \neq 0$. Then $\sigma_u\sigma_w$ and $\sigma_w\sigma_v$ are half-turns, by Proposition 6.12, and $\sigma_u\sigma_v = (\sigma_u\sigma_w)(\sigma_w\sigma_v)$. □

Proposition 6.14. *If $\dim V \geq 3$ then the derived group of $\mathrm{O}(V)$ is equal to the derived group of $\mathrm{SO}(V)$.*

Proof. Write $\mathrm{O}'(V)$ and $\mathrm{SO}'(V)$ for the respective derived groups. If u and v are nonzero anisotropic vectors note that $[\sigma_u, \sigma_v] = (\sigma_u\sigma_v)^2$. Set

$$K = \langle(\sigma_u\sigma_v)^2 \colon 0 \neq u,\ v \text{ anisotropic}\rangle \leq \mathrm{O}'(V).$$

Since $\tau\sigma_u\tau^{-1} = \sigma_{\tau u}$ for any $\tau \in \mathrm{O}(V)$ it is clear that $K \lhd \mathrm{O}(V)$. By the Cartan-Dieudonné Theorem $\mathrm{O}(V)/K = \langle \sigma_u K \colon 0 \neq u$ anisotropic$\rangle$, and these generators of $\mathrm{O}(V)/K$ commute since K is generated by the commutators $[\sigma_u, \sigma_v]$. Thus $\mathrm{O}(V)/K$ is abelian, and hence $K \geq \mathrm{O}'(V)$.

Suppose first that n is odd, so that $-1_V \in \mathrm{O}(V) \setminus \mathrm{SO}(V)$. Consequently every $-1_V\sigma_u = -\sigma_u$ is in $\mathrm{SO}(V)$, $[-\sigma_u, -\sigma_v] = [\sigma_u, \sigma_v] = (\sigma_u\sigma_v)^2$, and $\mathrm{O}'(V) \leq \mathrm{SO}'(V)$ in that case.

Suppose next that n is even, so $n \geq 4$. Fix u and v as above and set $U = \langle u, v\rangle$. Then $U^\perp$ cannot be totally isotropic, or we would have $U^\perp \subseteq U^{\perp\perp} = U$, and then $\dim U^\perp = n - \dim U \geq n - 2 \geq 2$. It would follow that $U = U^\perp$, so U is totally isotropic, whereas u and v are nonzero anisotropic vectors in U, a contradiction. Thus we may choose $w \in U^\perp$ with $Q(w) \neq 0$. Then $\sigma_u\sigma_w\sigma_u^{-1} = \sigma_{\sigma_u w} = \sigma_w$, and $\sigma_u\sigma_w = \sigma_w\sigma_u$. Similarly $\sigma_v\sigma_w = \sigma_w\sigma_v$. But now $\sigma_u\sigma_w$ and $\sigma_v\sigma_w$ are rotations, and $[\sigma_u\sigma_w, \sigma_v\sigma_w] = [\sigma_u, \sigma_v]$, so again $\mathrm{O}'(V) \leq \mathrm{SO}'(V)$.

Clearly then $\mathrm{O}'(V) = \mathrm{SO}'(V)$. □

Proposition 6.15. $Z(\mathrm{O}(V)) = \{\pm 1_V\}$.

Proof. If $\dim V = 2$ and V is isotropic this is an easy consequence of Proposition 6.1, so we may assume that V is anisotropic if $n = 2$. Take $\tau \in Z = Z(\mathrm{O}(V))$ and let $u \in V$ be any nonzero anisotropic vector. Then $\sigma_u = \tau\sigma_u\tau^{-1} = \sigma_{\tau u}$, so $\tau u \in \langle u\rangle$, say $\tau u = au$. Then

$$Q(u) = Q(\tau u) = a^2 Q(u),$$

so $a = \pm 1$, $\tau u = \pm u$. Now let $u_1, \ldots, u_n$ be an orthogonal basis for V (necessarily anisotropic since B is nondegenerate), and say that $\tau u_i = \epsilon_i u_i$ for each i, $\epsilon_i = \pm 1$. Take $i \neq j$. If $u_i + u_j$ is anisotropic then

$$\tau(u_i + u_j) = \pm(u_i + u_j) = \epsilon_i u_i + \epsilon_j u_j,$$

so $\epsilon_i = \epsilon_j$. If $n = 2$ the proof is complete.

If $n \geq 3$ and if $u_i + u_j$ is isotropic choose $k \neq i$, $k \neq j$. Then

$$Q(u_i + u_j + u_k) = Q(u_i + u_j) + Q(u_k) = Q(u_k) \neq 0,$$

so

$$\tau(u_i + u_j + u_k) = \pm(u_i + u_j + u_k) = \epsilon_i u_i + \epsilon_j u_j + \epsilon_k u_k,$$

and again $\epsilon_i = \epsilon_j$. □

Note that $-1_V \in \mathrm{SO}(V)$ if and only if n is even.

We shall assume for the remainder of the chapter that the Witt index $m = m(V)$ is at least 1, i.e. that V is isotropic. When $n = \dim V = 2$ then

V is a hyperbolic plane, and $\mathrm{O}(V)$ was completely described in Proposition 6.1, so we shall also assume that $n \geq 3$.

Let $u \in V$ be isotropic, choose $y \in \langle u \rangle^\perp$, and define a linear transformation $\eta_{u,y}\colon \langle u \rangle^\perp \to \langle u \rangle^\perp$ by setting

$$\eta_{u,y}(w) = w + B(w, y)u.$$

Note that $\eta_{au,y} = \eta_{u,ay}$ if $a \in F$, and that $\eta_{u,0} = 1_{\langle u \rangle^\perp}$. If y, $z \in \langle u \rangle^\perp$ then

$$\eta_{u,y+z}(w) = w + B(w, y + z)u,$$

whereas

$$\begin{aligned} \eta_{u,z}\eta_{u,y}(w) &= \eta_{u,z}(w + B(w, y)u) \\ &= w + B(w, y)u + B(w + B(w + B(w, y)u, z))u \\ &= w + B(w, y)u + B(w, z)u \;=\; w + B(w, y + z)u, \end{aligned}$$

i.e. $\eta_{u,y+z} = \eta_{u,z}\eta_{u,y}$. In particular

$$\eta_{u,y}\eta_{u,-y} = \eta_{u,0} = 1,$$

and $\eta_{u,y} \in \mathrm{GL}(\langle u \rangle^\perp)$, with $\eta_{u,y}^{-1} = \eta_{u,-y}$. In fact, $\eta_{u,y} \in \mathrm{O}(\langle u \rangle^\perp)$, for if $w \in \langle u \rangle^\perp$ then

$$Q(\eta_{u,y}w) = B(w + B(w, y)u, w + B(w, y)u) = B(w, w) = Q(w)$$

(recall that u is isotropic).

Proposition 6.16. *If $u \in V$ is isotropic and $y \in \langle u \rangle^\perp$ then the isometry $\eta_{u,y}$ of $\langle u \rangle^\perp$ extends uniquely to an isometry $\rho_{u,y} \in \mathrm{O}(V)$.*

Proof. By Witt's Extension Theorem (5.2) $\eta_{u,y}$ extends to some $\rho_{u,y} \in \mathrm{O}(V)$. Choose $v \in V$ so that (u, v) is a hyperbolic pair (e.g. see the proof of Proposition 4.12), and set $H = \langle u,\, v \rangle$. Thus $V = H \bigoplus H^\perp$; note that $H^\perp \subseteq \langle u \rangle^\perp$. If $\tau \in \mathrm{O}(V)$ is any extension of $\eta_{u,y}$ to V then $\tau^{-1}\rho_{u,y}|_{\langle u \rangle^\perp} = 1_{\langle u \rangle^\perp}$, so $\tau^{-1}\rho_{u,y}|_{H^\perp} = 1_{H^\perp}$. Since $\tau^{-1}\rho_{u,y}$ fixes the isotropic vector $u \in H$ we also have $\tau^{-1}\rho_{u,y}|_H = 1_H$ by the exercise on page 40, so $\tau^{-1}\rho_{u,y} = 1_V$, and $\tau = \rho_{u,y}$. □

Corollary 6.17. *If y, $z \in \langle u \rangle^\perp$ then $\rho_{u,y+z} = \rho_{u,z}\rho_{u,y}$.*

Proof. Both $\rho_{u,y+z}$ and $\rho_{u,z}\rho_{u,y}$ extend $\eta_{u,y+z}$ to V; apply the uniqueness in the Proposition. □

Corollary 6.18. *If $a \in F^\star$ then $\rho_{au,y} = \rho_{u,ay}$.*

Corollary 6.19. $\rho_{u,y}^{-1} = \rho_{u,-y}$.

Corollary 6.20. $\rho_{u,y} = 1_V$ *if and only if* $y \in \langle u \rangle = \mathrm{rad}\langle u \rangle^\perp$.

Corollary 6.21. *If $\tau \in \mathrm{O}(V)$ then $\tau\rho_{u,y}\tau^{-1} = \rho_{\tau u,\tau y}$.*

Proof. If $w \in \langle u\rangle^\perp$ then

$$\begin{aligned}\tau\rho_{u,y}\tau^{-1}(\tau w) &= \tau\rho_{u,y}w = \tau\eta_{u,y}w \\ &= \tau(w + B(w,y)u) \\ &= \tau w + B(\tau w, \tau y)\tau u \\ &= \eta_{\tau w,\tau y}(\tau w) = \rho_{\tau w,\tau y}(\tau w).\end{aligned}$$

Thus $\tau\rho_{u,y}\tau^{-1}$ and $\rho_{\tau w,\tau y}$ both extend $\eta_{\tau u,\tau y}$ from $\tau\langle u\rangle^\perp$ to all of V, and again we apply uniqueness. □

The transformations $\rho_{u,y}$ are often called *Siegel transformations*, having been introduced by Siegel in [**63**].

Exercise

Show that the extension of $\eta_{u,y}$ to $\rho_{u,y}$ can be given explicitly as follows. Take (u,v) and $H = \langle u,\ v\rangle$ as in the proof of Proposition 6.16. Show first that we may assume $y \in H^\perp$ (replace y by $y_1 \in H^\perp$ with $\eta_{u,y_1} = \eta_{u,y}$). Then $\rho_{u,y} = \eta_{u,y}$ on $\langle u\rangle^\perp$ and $\rho_{u,y}(v) = v - \frac{1}{2}\,Q(y)\,u - y$.

Proposition 6.22. *Suppose that (u,v) is a hyperbolic pair in V and that H is the hyperbolic plane $\langle u,\ v\rangle$, and set $K_u = \{\rho_{u,y} \colon y \in H^\perp\}$. Then $K_u \lhd \mathrm{Stab}_{\mathrm{O}(V)}(u)$ and the map $\varphi\colon y \mapsto \rho_{u,y}$ is an isomorphism from the additive group $H^\perp$ to K_u. In particular K_u is abelian.*

Proof. The elements of K_u stabilize u since u is isotropic, and $K_u \lhd \mathrm{Stab}(u)$ by Corollary 6.21. By Corollary 6.17 φ is a homomorphism. If $y \in \operatorname{Ker}\varphi$ then $B(w,y) = 0$, all $w \in \langle u\rangle^\perp$, so $y \in \mathrm{rad}(\langle u\rangle^\perp) \cap H^\perp = \langle u\rangle \cap H^\perp = 0$, and φ is injective. □

Remark

By the exercise above the subgroup K_u in Proposition 6.22 depends only on u, and not on the choice of v for which (u,v) is a hyperbolic pair.

Exercise

Show that $(\rho_{u,y} - 1_V)^3 = 0$ in all cases, hence that $\rho_{u,y} \in \mathrm{SO}(V)$, and so $K_u \le \mathrm{SO}(V)$. (In fact, if $\tau - 1$ is nilpotent, where τ is any linear transformation, then $\tau \in \mathrm{SL}(V)$.)

Next: an important subgroup of $\mathrm{SO}(V)$. Define

$$\Omega = \Omega(V) = \langle \cup K_u \colon u \text{ is isotropic in } V\rangle \le \mathrm{SO}(V).$$

If $\tau \in \mathrm{O}(V)$ then by Corollary 6.21 we see that $\tau K_u \tau^{-1} = K_{\tau u}$, and hence $\Omega \lhd \mathrm{O}(V)$.

Proposition 6.23. *Suppose G is a group generated by a set S of involutions. Then $\tau^2 \in G'$ for all $\tau \in G$.*

Proof. Write $\tau = \sigma_1 \cdots \sigma_k$, each $\sigma_i \in S$ (so $\sigma_i^2 = 1$). Then

$$\tau^2 G' = \sigma_1 \cdots \sigma_k \sigma_1 \cdots \sigma_k G' = \sigma_1^2 \cdots \sigma_k^2 G' = G'$$

(since G/G' is abelian), and $\tau^2 \in G'$. □

Proposition 6.24. $\Omega \leq \mathrm{O}'(V)$.

Proof. Let $\rho_{u,y}$ be one of the generators for Ω. Then $\rho_{u,y} = \rho_{u,y/2}^2$ is a square in $\mathrm{O}(V)$. Since $\mathrm{O}(V)$ is generated by reflections, $\rho_{u,y} \in \mathrm{O}'(V)$ by Proposition 6.23, hence $\Omega \leq \mathrm{O}'(V)$. □

Set $\mathcal{C} = \{u \in V \colon Q(v) = 0\}$. Note that if $\widehat{B} = \mathrm{diag}(b_1, \ldots, b_n)$ relative to a basis $\{v_1, \ldots, v_n\}$ for V then $\mathcal{C} = \{\sum_i x_i v_i \colon \sum_i b_i x_i^2 = 0\}$; $\mathcal{C}$ is called a *quadric cone* in V. Clearly $\mathrm{O}(V)$ acts as a permutation group on $\mathcal{C}$. Let $\mathbb{P}\mathcal{C} = \{[u] \in \mathbb{P}_{n-1}(V) \colon u \in \mathcal{C}\}$, a *projective quadric cone.* There is also an obvious permutation action of $\mathrm{O}(V)$ on $\mathbb{P}\mathcal{C}$.

Proposition 6.25. *The kernel of the action of $\mathrm{O}(V)$ on $\mathbb{P}\mathcal{C}$ is $Z = \{\pm 1\}$.*

Proof. If K denotes the kernel it is clear that $Z \leq K$. Take $\tau \in K$. Let (u, v) be a hyperbolic pair in V, choose anisotropic $w \neq 0$ in $\langle u, v\rangle^\perp$, and set $x = w + v - \frac{1}{2}Q(w)u$. Then $Q(x) = Q(w) - Q(w) = 0$, so x is isotropic, and $[x] \in \mathbb{P}\mathcal{C}$. Say $\tau u = au$, $\tau v = bv$, and $\tau x = cx$, $a, b, c \in F^\star$. Then

$$cx = cw + cv - \frac{1}{2}cQ(w)u = \tau x = \tau w + bv - \frac{1}{2}Q(w)au.$$

Since $\tau\langle u, v\rangle = \langle u, v\rangle$ and $w \in \langle u, v\rangle^\perp$ we conclude that $\tau w = cw$, and also $b = c$, $a = c$. Since $\langle u, v\rangle^\perp$ has a basis of vectors such as w we must have $\tau = c \cdot 1$, and $c = \pm 1$ since $\tau \in \mathrm{O}(V)$. □

If $0 \neq u \in V$ write $T_u = \mathcal{C} \cap \langle u\rangle^\perp$, the "trace" of the quadric cone $\mathcal{C}$ in the hyperplane $\langle u\rangle^\perp$. Write $\mathbb{P}T_u$ for the corresponding set in $\mathbb{P}_{n-1}(V)$, i.e.

$$\mathbb{P}T_u = \{[v] \in \mathbb{P}(V) \colon 0 \neq v \in T_u\}.$$

Proposition 6.26. *If u is isotropic in V then the group K_u is transitive on $\mathbb{P}\mathcal{C} \setminus \mathbb{P}T_u$.*

Proof. Choose $[v]$, $[w] \in \mathbb{P}\mathcal{C} \setminus \mathbb{P}T_u$, so $0 \neq v, w \in \mathcal{C}$ with $v, w \not\perp u$. It will suffice to find $\rho_{u,y} \in K_u$ such that $\rho_{u,y}(v) \in \langle w\rangle$. We may assume that $B(u, v) = B(u, w) = 1$. Thus (u, v) is a hyperbolic pair, and we have $V = \langle u, v\rangle \oplus \langle u, v\rangle^\perp$. Write $w = au + bv + z$, with $z \in \langle u, v\rangle^\perp$. Then $1 = B(u, w) = b$, and $0 = Q(w) = 2a + Q(z)$, so $w = v - \frac{1}{2}Q(z)u + z = \rho_{u,-z}(v)$ by the first exercise on page 53. □

If (u, v) is a hyperbolic pair in V call the ordered pair $([u], [v])$ a *(projective) hyperbolic pair* in $\mathbb{P}_{n-1}(V)$. Note that then $[u]$, $[v] \in \mathbb{P}\mathcal{C}$, and they are distinct.

Proposition 6.27. *The group Ω is transitive on $\mathbb{P}\mathcal{C}$ and on the set of hyperbolic pairs in $\mathbb{P}\mathcal{C}$.*

Proof. Suppose $[w] \neq [v]$ in $\mathbb{P}\mathcal{C}$. We show first that there exists $[u] \in \mathbb{P}\mathcal{C}$ such that $([u], [w])$ and $([u], [v])$ are projective hyperbolic pairs. Suppose first that $([w], [v])$ is a hyperbolic pair; then we may assume that (u, v) is a hyperbolic pair in V. Choose $x \neq 0$ in $\langle u, v\rangle^{\perp}$, x anisotropic, and set

$$u = w - \frac{1}{2}Q(x)\, v + x \neq 0.$$

Observe then that $Q(u) = 0$, so $[u] \in \mathbb{P}\mathcal{C}$. Also $B(u, w) = -\frac{1}{2}Q(x) \neq 0$, and $B(u, v) = 1$, so $([u], [w])$ and $([u], [v])$ are hyperbolic pairs.

Suppose next that $B(w, v) = 0$. Choose $f \in V^*$ with $f(w) = f(v) = 1$, and hence obtain $u \in V$ (by Corollary 2.2) with $B(u, w) = B(u, v) = 1$. By replacing u by $u + aw$ for appropriately chosen $a \in F$ we may assume that $Q(u) = 0$, so $[u] \in \mathbb{P}\mathcal{C}$.

With $[u]$ so chosen we may choose $\rho_{u,y} \in K_u \leq \Omega$, by Proposition 6.26, with $\rho_{u,y}[w] = [v]$, and Ω is transitive on $\mathbb{P}\mathcal{C}$.

Next let $([u_1], [v_1])$ and $([u_2], [v_2])$ be distinct hyperbolic pairs in $\mathbb{P}\mathcal{C}$, and assume that (u_1, v_1) and (u_2, v_2) are hyperbolic pairs in V. Choose $\tau \in \Omega$ such that $\tau[u_1] = [u_2]$, say $\tau u_1 = bu_2$, $b \neq 0$ in F. Then

$$B(u_2, \tau v_1) = B(\tau^{-1}u_2, v_1) = b^{-1}B(u_1, v_1) = b^{-1} \neq 0,$$

so $([u_2], [\tau v_1])$ is a projective hyperbolic pair. By Proposition 6.26 there exists $\sigma \in K_{u_2} \leq \Omega$ such that $\sigma[u_2] = [u_2]$ and $\sigma[\tau v_1] = [v_2]$. Then $\sigma\tau$ maps $([u_1], [v_1])$ to $([u_2], [v_2])$. □

Proposition 6.28. $\Omega = \mathrm{O}'(V) = \mathrm{SO}'(V)$.

Proof. Let (u, v) be a hyperbolic pair in V, and set $H = \langle u, v\rangle$. If x is any nonzero anisotropic vector in V we show first that $\tau x \in H$ for some $\tau \in \Omega$. Set $x_1 = u + \frac{1}{2}Q(x)v$ and note that $Q(x_1) = Q(x)$. Thus $\langle x_1\rangle$ and $\langle x\rangle$ are isometric, and by the Witt Extension Theorem (5.2) there exists $\sigma \in \mathrm{O}(V)$ with $\sigma x_1 = x$. By Proposition 6.27 there exists $\tau \in \Omega$ with $\tau([\sigma u]) = [u]$ and $\tau([\sigma v]) = [v]$. Since $x_1 \in H$, $x = \sigma x_1$ is in H, and therefore $\tau x \in \tau\sigma H = \tau\sigma\langle u, v\rangle = \tau(\sigma\langle u\rangle + \sigma\langle v\rangle) = \langle u\rangle + \langle v\rangle = H$, as needed.

A consequence: for any nonzero anisotropic x there exists $\tau \in \Omega$ such that $\tau\sigma_x\tau^{-1} = \sigma_{\tau x}$, with $\tau x \in H$, i.e. any reflection can be conjugated to a reflection along a vector in H by means of a transformation in Ω.

Write $\mathrm{O}_H(V)$ for $\langle \sigma_w \colon 0 \neq w \in H,\ w$ anisotropic$\rangle$. Clearly $\mathrm{O}_H(V) \cong \mathrm{O}(H)$; each $\tau \in \mathrm{O}(H)$ extends trivially to $\tau \in \mathrm{O}_H(V)$ via $\tau|_{H^\perp} = 1_{H^\perp}$. Write $\mathrm{SO}_H(V)$ for $\mathrm{O}_H(V) \cap \mathrm{SO}(V)$, clearly isomorphic with $\mathrm{SO}(H)$.

If $\sigma \in \mathrm{SO}(V)$ write $\sigma = \sigma_{x_1} \cdots \sigma_{x_{2k}}$, each x_i nonzero and anisotropic. Choose $\tau_i \in \Omega$ such that $\tau_i x_i \in H$, and hence

$$\sigma = (\tau_1^{-1} \sigma_{\tau_1 x_1} \tau_1) \cdots (\tau_{2k}^{-1} \sigma_{\tau_{2k} x_{2k}} \tau_{2k}).$$

Since $\Omega \triangleleft \mathrm{O}(V)$ it follows that all the $\sigma_{\tau_i x_i}$ can be moved to the right to obtain

$$\sigma = \tau \sigma_{\tau_1 x_1} \cdots \sigma_{\tau_{2k} x_{2k}},$$

with $\tau \in \Omega$, and hence $\mathrm{SO}(V) \leq \Omega\,\mathrm{SO}_H(V)$. But $\Omega \leq \mathrm{SO}(V)$, so in fact $\mathrm{SO}(V) = \Omega\,\mathrm{SO}_H(V)$. Thus

$$\mathrm{SO}(V)/\Omega = \Omega\,\mathrm{SO}_H(V)/\Omega \cong \mathrm{SO}_H(V)/(\mathrm{SO}_H(V) \cap \Omega).$$

But $\mathrm{SO}_H(V) \cong \mathrm{SO}(H) \cong F^\star$ is abelian. It follows that $\Omega \geq \mathrm{SO}'(V) = \mathrm{O}'(V)$ (Proposition 6.14), and hence $\Omega = \mathrm{O}'(V)$. □

As is customary we shall continue to write Ω, or $\Omega(V)$, rather than $\mathrm{O}'(V)$. We also continue to assume that V is a quadratic space, with $\dim V = n \geq 3$ and Witt index $m = m(V) > 0$.

Proposition 6.29. *If $n \neq 4$, or if $m \neq 2$, then Ω acts primitively on $\mathbb{P}\mathcal{C}$.*

Proof. This is easy if $m = 1$, for then *every* ordered pair $([u], [v])$ of distinct points from $\mathbb{P}\mathcal{C}$ is a (projective) hyperbolic pair. Thus Ω is doubly transitive on $\mathbb{P}\mathcal{C}$ by Proposition 6.27, and hence is primitive by Proposition 0.2.

Assume next that $m(V) \geq 2$, and hence also that $n \geq 5$. Consider the possibility of a block $\mathcal{B}$ in $\mathbb{P}\mathcal{C}$, i.e. suppose that $\mathcal{B} \subseteq \mathbb{P}\mathcal{C}$, with $|\mathcal{B}| > 1$ and either $\sigma\mathcal{B} = \mathcal{B}$ or $\sigma\mathcal{B} \cap \mathcal{B} = \emptyset$ for all $\sigma \in \Omega$. We must show that $\mathcal{B} = \mathbb{P}\mathcal{C}$.

Case 1: There exist $[x]$, $[y]$, distinct, in $\mathcal{B}$ with $x \perp y$. By Corollary 2.2 there exists $z \in V$ with $B(x, z) = 1$, $B(y, z) = 0$. Replacing z by $z + ax$ for appropriately chosen $a \in F$ we may further assume that $Q(z) = 0$, i.e. that z is isotropic. Set $H = \langle x,\ z\rangle$, so $V = H \perp H^\perp$, and note that $\dim H^\perp \geq 3$ and $H^\perp$ is nondegenerate. Also $y \in H^\perp$, so there exists $w \in H^\perp$ such that (y, w) is a hyperbolic pair. Now Proposition 6.26 applies to $H^\perp$, and there exists $\tau \in \Omega(H^\perp)$ such that $\tau[y] = [w]$. We may (and shall) view τ, however, as being in $\Omega(V)$, since it is a product of various $\rho_{u,v}$, for $u,\ v \in H^\perp \subseteq V$. Note that $\tau[x] = [x]$, since $x \in H = H^{\perp\perp}$. But now $[x] \in \mathcal{B} \cap \tau\mathcal{B}$, so $\tau\mathcal{B} = \mathcal{B}$, and hence $[w] = \tau[y] \in \mathcal{B}$. Consequently $([y], [w])$ is a hyperbolic pair in $\mathcal{B}$. This reduces Case 1 to the next case.

Case 2: There exist $[x]$, $[y]$ distinct in $\mathcal{B}$ such that $([x], [y])$ is a projective hyperbolic pair. Choose $[z] \neq [x]$ in $\mathbb{P}\mathcal{C}$. Then choose $[w] \in \mathbb{P}\mathcal{C}$ such that

$([x],[w])$ and $([z],[w])$ are hyperbolic pairs (see e.g. the proof of Proposition 6.27). By Proposition 6.27 there exists $\tau \in \Omega$ such that $\tau[x] = [x]$ and $\tau[y] = [w]$. Then $[x] = \tau[x] \in \mathcal{B} \cap \tau\mathcal{B}$, so $\tau\mathcal{B} = \mathcal{B}$, and hence $[w] = \tau[y] \in \mathcal{B}$. Thus $[w] = \sigma[w] \in \mathcal{B} \cap \sigma\mathcal{B}$, so $\sigma\mathcal{B} = \mathcal{B}$, and hence $[z] = \sigma[x] \in \mathcal{B}$. Since $[z]$ was arbitrary $\mathcal{B} = \mathbb{P}\mathcal{C}$. □

Proposition 6.30. *Suppose V is a quadratic space with $n = \dim V \geq 3$ and $m = m(V) > 0$. Then $\Omega' = \Omega$ unless* (i) *$n = 4$ and $m = 2$, or* (ii) *$n = 3$ and $|F| = 3$.*

Proof. We assume throughout that $(n, m) \neq (4, 2)$. It will suffice to show that if u is isotropic and $y \perp u$ then $\rho_{u,y} \in \Omega'$. Choose v such that (u, v) is a hyperbolic pair, set $H = \langle u, v \rangle$, and define $\mathrm{O}_H(V) \leq \mathrm{O}(V)$ as in the proof of Proposition 6.28, so $\mathrm{O}_H(V) \cong \mathrm{O}(H)$. Recall from the first exercise on page 53 that we may assume $y \in H^\perp$. By Proposition 6.1 there exists $\tau \in \mathrm{O}_H(V)$ with $\tau u = v$ and $\tau v = u$ (note that $\tau^2 = 1$), and if $a \in F^\star$ there exists $\eta_a \in \mathrm{O}_H(V)$ such that $\eta_a u = au$, $\eta_a v = a^{-1}v$. An easy calculation shows that $[\tau, \eta_a] = \eta_{a^2} \in \mathrm{O}'(V) = \Omega$. Now use the corollaries to Proposition 6.16 to calculate:

$$\eta_{a^2}\rho_{u,y}\eta_{a^2}^{-1}\rho_{u,-y} = \rho_{a^2u,y}\rho_{u,-y} = \rho_{u,a^2y}\rho_{u,-y} = \rho_{u,(a^2-1)y} \in \Omega'.$$

Case 1. $|F| \neq 3$ (hence $|F| \geq 5$). Choose $a \in F^\star$, $a^2 \neq 1$. Set $y_1 = y/(a^2-1)$, and apply the calculation above to ρ_{u,y_1} to obtain $\rho_{u,y_1} \in \Omega'$.

Case 2. $|F| = 3$. Thus $n \geq 4$, and if $n = 4$ then $m \neq 2$, so $m = 1$. In considering $\rho_{u,y}$ we may continue to assume that $y \in H^\perp$. Since $H^\perp$ has an orthogonal basis every $y \in H^\perp$ is a sum of anisotropic vectors. Since $\rho_{u,y+z} = \rho_{u,y}\rho_{u,z}$ it will suffice, then, to prove that $\rho_{u,y} \in \Omega'$ for all anisotropic vectors y. We show that if $0 \neq y$ is anisotropic in $H^\perp$ then there exists $z \in H^\perp$ such that $z \perp y$ and $Q(z) = Q(y)$.

Suppose first that $n = 4$, so there exists $z \in H^\perp$ such that $\{y, z\}$ is an orthogonal basis. Since H is isotropic, and $m = 1$, $H^\perp$ must be anisotropic. But then $Q(y) = -Q(z)$ is impossible, since then $y + z$ would be isotropic. Thus $Q(y) = Q(z) = \pm 1$.

Suppose next that $n > 4$, so $\dim H^\perp \geq 3$, and again take $y \neq 0$ in $H^\perp$, y anisotropic. Then $\langle y \rangle^\perp \cap H^\perp$ is the orthogonal complement of $\langle y \rangle$ in $H^\perp$; it has dimension at least 2 and is nondegenerate since $\langle y \rangle$ is nondegenerate. Thus $Q|_{\langle y \rangle^\perp \cap H^\perp}$ is universal by Proposition 4.8, and there exists $z \in \langle y \rangle^\perp \cap H^\perp$ (and hence $z \perp y$) such that $Q(z) = Q(y)$. By Witt's Extension Theorem (5.2) we may choose $\tau \in \mathrm{O}(V)$ with $\tau u = u$, $\tau y = -z$, and $\tau z = y$. Thus $\tau^2 u = u$, $\tau^2 y = -y$, and $\tau^2 z = -z$, and $\tau^2 \in \mathrm{O}'(V) = \Omega$ by Proposition 6.23.

Finally,

$$\tau^2 \rho_{u,y} \tau^{-2} \rho_{u,-y} = \rho_{u,-y} \rho_{u,-y} = \rho_{u,-2y} = \rho_{u,y} \in \Omega'.$$

□

By Proposition 6.25 the kernel of the action of Ω on the projective cone $\mathbb{P}\mathcal{C}$ is $\Omega \cap \{\pm 1\}$, so the kernel is either ± 1 (if $-1 \in \Omega$) or 1 (if $-1 \notin \Omega$). (Note that $-1 \notin \Omega$ if n is odd, since $\Omega \leq \mathrm{SO}(V)$.)

It is nontrivial to determine precisely when $-1 \in \Omega$ – we postpone the question until later. At any rate we define the corresponding projective group as usual:

$$\mathrm{P}\Omega(V) = \Omega(V)/(\Omega(V) \cap \{\pm 1\}).$$

Thus $\mathrm{P}\Omega(V)$ acts faithfully on $\mathbb{P}\mathcal{C}$.

Theorem 6.31 (Dickson-Dieudonné). *Suppose that V is a quadratic space with $\dim V = n \geq 3$ and Witt index $m > 0$. Assume that $(n, m) \neq (4, 2)$, and if $n = 3$ assume that $|F| \neq 3$. Then $\mathrm{P}\Omega(V)$ is simple.*

Proof. $\mathrm{P}\Omega$ acts faithfully on $\mathbb{P}\mathcal{C}$, and the action is primitive by Proposition 6.29. If $[u] \in \mathbb{P}\mathcal{C}$ then $K_u \lhd \mathrm{Stab}_{\mathrm{P}\Omega}([u])$, and K_u is abelian by Proposition 6.22. If $\tau \in \Omega$ then $\tau K_u \tau^{-1} = K_{\tau u}$, and τu ranges over all of $\mathbb{P}\mathcal{C}$ since Ω is transitive on $\mathbb{P}\mathcal{C}$ (Proposition 6.27). Thus $\langle \cup \tau K_u \tau^{-1} \colon \tau \in \Omega \rangle = \Omega$. Since $\Omega' = \Omega$ (Proposition 6.28), $\mathrm{P}\Omega$ is simple by Iwasawa's Theorem (0.5). □

The identification with groups of Chevalley type is a bit more complicated than for earlier cases. If $n = 2m+1$ is odd then $\mathrm{P}\Omega(V)$ is the Chevalley group $B_m(F)$. If $n = 2m$ is even, and if V is hyperbolic, then $\mathrm{P}\Omega(V)$ is the Chevalley group $D_m(F)$. If $n = 2m$ but V is not hyperbolic, then $\mathrm{P}\Omega(V)$ is the twisted group ${}^2D_m(K)$, where K is a degree 2 field extension of F, and $V^K = K \otimes_F V$ has Witt index m. See Carter [**11**], pages 184 and 271, for details.

The orders of the groups under discussion (over finite fields) will be calculated in Chapter 9.

Chapter 7

O(V), V Euclidean

For students of physics and vector calculus the most familiar of the classical groups are $\mathrm{O}(\mathbb{R}^2)$ and $\mathrm{O}(\mathbb{R}^3)$, the groups of isometries of 2- and 3-dimensional Euclidean space. It is perhaps of interest to have a close look at these groups, and more generally at $\mathrm{O}(\mathbb{R}^n)$, even though it is not central to the goals of the text.

Thus for this chapter we take $F = \mathbb{R}$ and take V to be ordinary n-dimensional real Euclidean space, i.e. the bilinear form B is the usual inner product: $B(u, v) = u^t v$ and $\widehat{B} = I$, relative to an orthonormal basis.

The results of the chapter will not be used in subsequent chapters.

As is customary we will write $\|v\|^2 = Q(v)$, so the *norm* or *length* of $v \in V$ is $\|v\| = \sqrt{B(v, v)} = \sqrt{v^t v}$, all $v \in V$. We define the *distance* between u and v in V to be $d(u, v) = \|u - v\|$. Denote by $\mathbb{S}$ the *unit sphere* in V. When dimension is an issue we write $\mathbb{S} = \mathbb{S}_{n-1}$ if dim $V = n$, since $\mathbb{S}$ is a surface of codimension 1.

If $n = 2$ then $\mathrm{SO}(V) = \mathrm{SO}(2, \mathbb{R})$, which consists of all rotations

$$\rho_\theta = \begin{bmatrix} \cos\theta & -\sin\theta \\ \sin\theta & \cos\theta \end{bmatrix}, \quad \theta \in \mathbb{R},$$

and the map $\rho\colon \theta \mapsto \rho_\theta$ is a homomorphism from the additive group $\mathbb{R}$ onto $\mathrm{SO}(2, \mathbb{R})$, with kernel $\langle 2\pi \rangle$. Thus $\mathrm{SO}(V) \cong \mathbb{R}/\langle 2\pi \rangle$, which is also isomorphic with the (multiplicative) group of all complex numbers of absolute value 1, commonly known as the *circle group*. Consequently $\mathrm{SO}(V)$ is abelian, and $\mathrm{O}(V)$ is a semidirect product $\mathrm{SO}(V) \rtimes \langle \sigma_u \rangle$ for any reflection σ_u.

Proposition 7.1. *If* $n = 2$ *then* $\mathrm{SO}(V)$ *is transitive on the unit sphere (circle)* $\mathbb{S} = \mathbb{S}_1$.

Proof. We may choose an orthogonal basis $\{e_1, e_2\} \subseteq \mathbb{S}$ for V, and hence assume that $V = \mathbb{R}^2$ (column vectors), and represent $\mathrm{O}(V)$ as $\mathrm{O}(2, \mathbb{R})$. If $u, v \in \mathbb{S}$ we may write

$$u = \begin{bmatrix} \cos\alpha \\ \sin\alpha \end{bmatrix} \quad \text{and} \quad v = \begin{bmatrix} \cos\beta \\ \sin\beta \end{bmatrix}$$

for appropriate α and β. Then

$$\rho_{\beta-\alpha} = \begin{bmatrix} \cos(\beta-\alpha) & -\sin(\beta-\alpha) \\ \sin(\beta-\alpha) & \cos(\beta-\alpha) \end{bmatrix} \in \mathrm{SO}(2, \mathbb{R}),$$

and $\rho_{\beta-\alpha} u = v$ by elementary trigonometric identities. □

Assume for the rest of the chapter that $\dim V = n \geq 3$.

Proposition 7.2. $\mathrm{SO}(V)$ *is transitive on the unit sphere* $\mathbb{S} = \mathbb{S}_{n-1}$.

Proof. Take $u, v \in \mathbb{S}$, $u \neq v$. If $u = -v$ choose $w \in V$ with $w \perp u$ and let $\sigma \in \mathrm{SO}(V)$ be the half-turn with axis $\langle u, w\rangle^{\perp}$. Then $\sigma u = -u = v$. If $u \neq -v$ then $\langle u, v\rangle = W$ is a Euclidean plane. If we write $\mathbb{S}_1$ for $\mathbb{S} \cap W$ then $\mathbb{S}_1$ is the unit sphere in W, and by Proposition 7.1 there exists $\sigma \in \mathrm{SO}(W)$ with $\sigma u = v$. But then $\tau = \sigma \bigoplus 1_{W^{\perp}} \in \mathrm{SO}(V)$ and $\tau u = v$. □

If (u_1, v_1) and (u_2, v_2) are ordered pairs of points in V with $d(u_1, v_1) = d(u_2, v_2)$ we call them *equidistant* pairs of points.

Proposition 7.3. $\mathrm{SO}(V)$ *acts transitively on equidistant pairs of points in the unit sphere* $\mathbb{S}$.

Proof. Suppose (u_1, v_1) and (u_2, v_2) are equidistant. First choose $\sigma \in \mathrm{SO}(V)$ so that $\sigma u_1 = u_2$. Then

$$d(u_2, \sigma v_1) = d(\sigma u_1, \sigma v_1) = d(u_1, v_1) = d(u_2, v_2),$$

so $(u_2, \sigma v_1)$ and (u_2, v_2) are equidistant. We need next $\tau \in \mathrm{Stab}_{\mathrm{SO}(V)}(u_2)$ such that $\tau\sigma v_1 = v_2$. Note that

$$d(u_2, v_2) = 2 - 2B(u_2, v_2) = 2 - 2B(u_2, \sigma v_1),$$

so $B(u_2, v_2) = B(u_2, \sigma v_1)$. If $B(u_2, v_2) = 0$ set $w_2 = v_2$ and $w_1 = \sigma v_1$, otherwise set

$$w_2' = u_2 - v_2/B(u_2, v_2), \quad w_1' = u_2 - \sigma v_1/B(u_2, \sigma v_1),$$

and then set

$$w_1 = w_1'/\|w_1'\|, \quad w_2 = w_2'/\|w_2'\| \in S.$$

In either case $w_1 \perp u_2$ and $w_2 \perp u_2$. Set $W = \langle u_2, w_1, w_2 \rangle$, $U = \langle w_1, w_2 \rangle$, so $W = \langle u_2 \rangle \bigoplus U$. By Proposition 7.1 there exists $\tau_1 \in \mathrm{SO}(U)$ with $\tau_1 w_1 = w_2$. Set $\tau_2 = \tau_1 \bigoplus 1_{\langle u_2 \rangle}$, and note that

$$\tau_2 \sigma v_1 = B(u_2, \sigma v_1)\tau_2(u_2 - w_1) = B(u_2, \sigma v_1)(u_2 - w_2) = v_2,$$

and $\tau_2 u_2 = u_2$. Set $\tau = \tau_2 \bigoplus 1_{W^\perp}$. Then $\tau \in \mathrm{SO}(V)$ and $\tau\sigma$ maps (u_1, v_1) to (u_2, v_2). □

Theorem 7.4. *If* $n = 3$ *then* SO(V) *is simple.*

Proof. Suppose $1 \neq K \lhd \mathrm{SO}(V)$ and choose $\sigma \in K$, $\sigma \neq 1$. Then σ has an axis of rotation $\langle u \rangle$, with $\|u\| = 1$, by Euler's Theorem (Corollary 6.10). Choose $y \in \mathbb{S}_2 \cap \langle u^\perp \rangle$ ("on the equator"), and let $\delta = d(y, \sigma y) = \|y - \sigma y\|$. Choose a positive integer k with $a = 2\sin \pi/2k$ satisfying $0 < a \leq \delta$, and set $b = a/\delta$, so $0 < b \leq 1$. Then set $z = \sqrt{1 - b^2}\, u + by \in S$, and $z_1 = \sigma z$, and note that $\|z - z_1\| = b\|y - \sigma y\| = b\delta = a$. Choose y_1 on the equator between y and σy such that $\|y - y_1\| = a$. By Proposition 7.2 there exists $\tau \in \mathrm{SO}(V)$ such that $\tau z = y$ and $\tau z_1 = y_1$. Then $\sigma_1 = \tau\sigma\tau^{-1} \in K$ and $\sigma_1 y = \tau\sigma\tau^{-1} y = \tau\sigma z = \tau z_1 = y_1$.

Choose vectors $y_2, y_3, \ldots, y_k = -y$ in $S_2 \cap \langle u^\perp \rangle$ such that $\|y_i - y_{i-1}\| = a = 2\sin \pi/2k$ (the angle between y_{i-1} and y_i is π/k). Applying the argument above to y_{i-1}, $2 \leq i \leq k$, we obtain $\sigma_i \in K$ and $\sigma_i(y_{i-1}) = y_i$. Thus if $\rho = \sigma_k\sigma_{k-1}\cdots\sigma_2\sigma_1$ then $\rho \in K$ and $\rho y = -y$. It follows that $\rho^2 y = y$, and of course $\rho^2 u = u$, so $\rho^2|_{\langle u,y \rangle} = 1_{\langle u,y \rangle}$, and hence $\rho^2 = 1_V$ by Proposition 5.7. Thus ρ is an involution and $\rho \in \mathrm{SO}(V)$, so ρ is a half-turn in K.

If W_1, W_2 are any two (hyper-) planes in V they are both Euclidean, hence isometric, and so there exists $\eta \in \mathrm{O}(V)$ with $\eta W_1 = W_2$ (Witt's Extension Theorem (5.2)). We may assume that $\eta \in \mathrm{SO}(V)$ (if not replace η by $\eta\sigma_{u_1}$, σ_{u_1} being the reflection through W_1). It follows easily that η conjugates the half-turn with fixed plane W_1 to the half-turn with fixed plane W_2. Thus all half-turns are conjugate, and since $\rho \in K$ all half-turns are in K. Thus $K = \mathrm{SO}(V)$ by Proposition 6.13, and SO(V) is simple. □

Theorem 7.5. *If* $n \geq 5$ *then* PSO(V) *is simple.*

Proof. If n is odd then $\mathrm{PSO}(V) = \mathrm{SO}(V)$; if n is even then $\mathrm{PSO}(V) = \mathrm{SO}(V)/Z$, with $Z = \{\pm 1\}$. In general we must show that if $Z \leq K \lhd \mathrm{SO}(V)$ then $K = \mathrm{SO}(V)$. Choose $\sigma \in K \setminus Z$. Then σ cannot leave every line invariant in V (otherwise $\sigma u = \pm u$ for all $u \in \mathbb{S}_{n-1}$, and the usual argument then shows that $\sigma = \pm 1_V$). It follows that σ cannot leave every plane (i.e. 2-dimensional subspace) invariant, since all lines are intersections of planes. Choose a plane P in V such that $\sigma P \neq P$ and let τ be the half-turn $-1_P \bigoplus 1_{P^\perp}$. Then $\sigma\tau\sigma^{-1}$ is the half-turn $-1_{\sigma P} \bigoplus 1_{\sigma P^\perp}$, so $\sigma\tau\sigma^{-1} \neq \tau$. Thus $1 \neq \sigma\tau\sigma^{-1}\tau^{-1} = \rho \in K$, and $\rho x = x$ if $x \in (P + \sigma P)^\perp$. Since $n \geq 5$

and $\dim(P + \sigma P) \leq 4$ there exists $u \in (P + \sigma P)^{\perp}$, $u \neq 0$. Note that $\rho \notin Z$ ($\rho \neq 1$, but $\rho u = u$, so $\rho \neq -1$), so by the argument above there is a line $L = \langle v \rangle$ with $\rho L \neq L$, say $\rho v = w \notin \langle v \rangle$. Set $\eta = \sigma_v \sigma_u$ and observe that

$$\rho\eta\rho^{-1}\eta^{-1} = \rho\sigma_v\rho^{-1}\rho\sigma_u\rho^{-1}\eta^{-1} = \sigma_{\rho v}\sigma_{\rho u}\eta^{-1} = \sigma_w\sigma_u\sigma_u\sigma_v = \sigma_w\sigma_v \in K.$$

Write $\alpha = \sigma_w\sigma_v$ and note that $\alpha|_{\langle v,w\rangle^{\perp}} = 1_{\langle v,w\rangle^{\perp}}$. Let W be some subspace of dimension 3 with v, $w \in W$, and write $\alpha = \beta \bigoplus 1_{W^{\perp}}$ with $1 \neq \beta \in \mathrm{SO}(W)$. Since $\alpha \in K \lhd \mathrm{SO}(V)$ we may conjugate $\beta \bigoplus 1_{W^{\perp}}$ by each $\gamma \bigoplus 1_{W^{\perp}}$, $\gamma \in \mathrm{SO}(W)$; the subgroup $M \bigoplus 1_{W^{\perp}}$ generated by those conjugates is normal in $\mathrm{SO}(W) \bigoplus 1_{W^{\perp}}$, hence is equal to $\mathrm{SO}(W) \bigoplus 1_{W^{\perp}}$ by Theorem 7.4. Thus all $\delta \bigoplus 1_{W^{\perp}} \in K$, $\delta \in \mathrm{SO}(W)$. In particular there is a half-turn in K, and hence $K = \mathrm{SO}(V)$ as in the proof of Theorem 7.4. □

Theorem 7.6. *If $n = 4$ then $\mathrm{PSO}(V) \cong \mathrm{SO}(W) \times \mathrm{SO}(W)$, where W is Euclidean of dimension 3.*

Proof. (Sketch) Identify V with the algebra $\mathbb{H}$ of real quaternions (see e.g. [**23**]), viewed as a real vector space, and let W be the subspace $\langle i, j, k \rangle$ (sometimes called the *imaginary prime*). Note that the norm function N on V coincides with the Euclidean quadratic form Q. Set $U = \{u \in V : Q(u) = 1\}$, the (multiplicative) group of "unit quaternions". The elements of U can all be written in the form $u = \cos\alpha + w \cdot \sin\alpha$, with $w \in W$ and $Q(w) = 1$. Define a map $\rho\colon U \to \mathrm{GL}(W)$ as follows: if $u \in U$ and $v \in W$ then $\rho\colon u \mapsto \rho_u$, with $\rho_u v = uvu^{-1}$. Then $Q(\rho_u v) = N(uvu^{-1}) = N(v) = Q(v)$, so $\rho_u \in \mathrm{O}(W)$. One checks easily that in fact $\rho_u \in \mathrm{SO}(W)$, and that ρ is a homomorphism with kernel $Z(U) = \{\pm 1\}$. If we write $u = \cos\alpha + w\sin\alpha$ then ρ_u is a rotation through angle 2α about axis $\langle w \rangle$. Thus ρ maps U onto $\mathrm{SO}(W)$ by Euler's Theorem (Corollary 6.10), and $U/Z(U) \cong \mathrm{SO}(W)$.

If $r, s \in U$ define $\theta_{r,s}(v) = rvs^{-1}$, all $v \in V$. Then θ is a homomorphism from $U \times U$ into $\mathrm{O}(V)$; one checks that in fact $\theta_{r,s} \in \mathrm{SO}(V)$. If $(r, s) \in \operatorname{Ker}\theta$ then $rvs^{-1} = v$ for all $v \in V$; taking $v = 1$ shows that $r = s$, so $\operatorname{Ker}\theta = \{(r, r) : r \in Z(U)\} = \{(1, 1), (-1, -1)\}$. Given $\tau \in \mathrm{SO}(V)$ say that $\tau(1) = t \in \mathbb{H}$. Define σ via $\sigma v = tv$ (quaternion multiplication), all $v \in V$. Then $\sigma \in \mathrm{SO}(V)$, $\sigma^{-1}\tau \in \mathrm{SO}(V)$, and $\sigma^{-1}\tau(1) = 1$. Thus $\sigma^{-1}\tau$ restricts to a rotation of W, and $\sigma^{-1}\tau|_W = \rho_s$ for some $s \in U$. But then $\sigma^{-1}\tau v = svs^{-1}$ for all $v \in V$, and hence $\tau v = \sigma(svs^{-1}) = (ts)vs^{-1} = \theta_{r,s}v$ if we set $r = ts$. Thus θ maps $U \times U$ onto $\mathrm{SO}(V)$. If we write $\eta\colon \mathrm{SO}(V) \to \mathrm{PSO}(V)$ for the canonical quotient map we have a natural commutative diagram.

$$\begin{array}{ccc} U \times U & \xrightarrow{\rho \times \rho} & \mathrm{SO}(W) \times \mathrm{SO}(W) \\ {\scriptstyle\theta}\downarrow & & \downarrow{\scriptstyle\varphi} \\ \mathrm{SO}(V) & \xrightarrow[\eta]{} & \mathrm{PSO}(V) \end{array}$$

Since $|\operatorname{Ker}(\rho\times\rho)| = 4$, $|\operatorname{Ker}\theta| = 2$, and $|\operatorname{Ker}\eta| = 2$, it follows easily that the map φ represented by the vertical arrow on the right is an isomorphism. $\square$

It follows, of course, that $\mathrm{PSO}(4,\mathbb{R})$ is *not* simple.

Chapter 8

Clifford Algebras (char $F \neq 2$)

Clifford algebras are perhaps properly viewed as far-reaching generalizations of the classical 4-dimensional algebra $\mathbb{H}$ of Hamiltonian quaternions. Just as $\mathbb{H}$ was useful early on for the study of subgroups of $\mathrm{O}(\mathbb{R}^3)$ (see e.g. [**23**]), we shall see that Clifford algebras play a crucial role in the study of orthogonal groups in general.

Quaternions have further applications, e.g. in vector analysis, and indeed Clifford algebras have a remarkable variety of further applications, ranging from differential geometry and topology to partial differential equations. Chapter 15 of [**32**] gives a brief listing; also see e.g. [**54**] for further applications.

We assume throughout this chapter that F is a field of characteristic different from 2, and that V is a (nondegenerate) quadratic space over F, with bilinear form B and quadratic form Q. Much of what follows also holds at characteristic 2, as we shall see later. Also, the assumption of nondegeneracy is not really necessary for some of the results that follow.

Suppose that A is an (associative) F-algebra with 1. Say that A is *compatible* with the quadratic form Q on V if there is a linear transformation $\varphi\colon V \to A$ such that $\varphi(v)^2 = Q(v)\cdot 1_A$ for all $v \in V$. It will be convenient to identify F with $F \cdot 1_A \subseteq A$, and hence assume that $F \subseteq A$ as a subalgebra. Thus compatibility of A with Q requires that $\varphi(v)^2 = Q(v)$, all $v \in V$.

Proposition 8.1. *If A is compatible with Q via $\varphi\colon V \to A$ then*

$$B(u,v) = \frac{1}{2}[\varphi(u)\varphi(v) + \varphi(v)\varphi(u)],$$

all u, $v \in V$.

Proof. Calculate:

$$\begin{aligned} B(u,v) &= \frac{1}{2}[Q(u+v) - Q(u) - Q(v)] \\ &= \frac{1}{2}[\varphi(u+v)^2 - \varphi(u)^2 - \varphi(v)^2] \\ &= \frac{1}{2}[(\varphi u + \varphi v)^2 - \varphi(u)^2 - \varphi(v)^2] \\ &= \frac{1}{2}[\varphi(u)\varphi(v) + \varphi(v)\varphi(u)]. \end{aligned}$$

□

Corollary 8.2. *Suppose $u, v \in V$. Then $u \perp v$ if and only if φu and φv "anticommute" in A, i.e.*

$$(\varphi u)(\varphi v) = -(\varphi v)(\varphi u).$$

Proposition 8.3. *Suppose A is compatible with Q via φ and $0 \neq v \in V$. Then φv is a unit in A if and only if v is anisotropic in V.*

Proof. $\Leftarrow$: $0 \neq Q(v) = (\varphi v)^2$, so $\varphi(v)Q(v)^{-1} = (\varphi v)^{-1}$.
$\Rightarrow$: If φv is a unit in A then $Q(v)(\varphi v)^{-1} = (\varphi v)^2(\varphi v)^{-1} = \varphi v$, and so $Q(v) \neq 0$. □

Proposition 8.4. *If A is compatible with Q, and if $u \neq 0$ is anisotropic in V, then*

$$\varphi(\sigma_u v) = -(\varphi u)(\varphi v)(\varphi u)^{-1}$$

for all $v \in V$, i.e. the effect of reflection of V along u carries over to the negative of conjugation by φu in A.

Proof. Again calculate:

$$\begin{aligned} \varphi(\sigma_u v) &= \varphi v - \frac{2B(v,u)}{Q(u)}\varphi u \\ &= \varphi v - \frac{[(\varphi v)(\varphi u) + (\varphi u)(\varphi v)]}{(\varphi u)^2}\varphi u \\ &= -(\varphi u)(\varphi v)(\varphi u)^{-1}. \end{aligned}$$

□

The preparation above leads to an important definition.

A *Clifford algebra* for a quadratic space V is an associative algebra C (with 1) that is compatible with the quadratic form Q of V, via a map $\varepsilon\colon V \to C$, and such that given any algebra A compatible with V (i.e.

with Q) via $\varphi\colon V \to A$, then there exists a unique algebra homomorphism $\widehat{\varphi}\colon C \to A$ so that the diagram

$$\begin{array}{ccc} V & \xrightarrow{\varepsilon} & C \\ & {}_{\varphi}\searrow & \downarrow{\scriptstyle \widehat{\varphi}} \\ & & A \end{array}$$

is commutative.

Proposition 8.5. *If V has a Clifford algebra C then C is unique up to an F-algebra isomorphism.*

Proof. Suppose that C_1 is also a Clifford algebra compatible with Q via $\varepsilon_1\colon V \to C_1$. We have

$$\begin{array}{ccc} V & \xrightarrow{\varepsilon} & C \\ & {}_{\varepsilon_1}\searrow & \downarrow{\scriptstyle \widehat{\varepsilon}_1} \\ & & C_1 \end{array} \quad \text{and} \quad \begin{array}{ccc} V & \xrightarrow{\varepsilon_1} & C_1 \\ & {}_{\varepsilon}\searrow & \downarrow{\scriptstyle \widehat{\varepsilon}} \\ & & C \end{array}$$

or $\varepsilon_1 = \widehat{\varepsilon}_1\varepsilon$ and $\varepsilon = \widehat{\varepsilon}\varepsilon_1$. Thus $\varepsilon_1 = \widehat{\varepsilon}_1\widehat{\varepsilon}\varepsilon_1$ and $\varepsilon = \widehat{\varepsilon}\,\widehat{\varepsilon}_1\varepsilon$. But then we have

$$\begin{array}{ccc} V & \xrightarrow{\varepsilon} & C \\ & {}_{\varepsilon}\searrow & \downarrow{\scriptstyle \widehat{\varepsilon}\widehat{\varepsilon}_1} \\ & & C \end{array} \quad \text{and} \quad \begin{array}{ccc} V & \xrightarrow{\varepsilon} & C \\ & {}_{\varepsilon}\searrow & \downarrow{\scriptstyle 1_C} \\ & & C \end{array}$$

and by the uniqueness in the definition we conclude that $\widehat{\varepsilon}\,\widehat{\varepsilon}_1 = 1_C$, the identity map on C. Similarly $\widehat{\varepsilon}_1\widehat{\varepsilon}$ is the identity map on C_1, so $\widehat{\varepsilon}$ and $\widehat{\varepsilon}_1$ are a pair of inverse isomorphisms. □

In order to construct a Clifford algebra for a given quadratic space V we consider first the *tensor algebra* $T = T(V)$ of the vector space V (see e.g. [**45**], page 139). Set $T_0 = F$, $T_1 = V$, $T_2 = V \otimes V$, and in general

$$T_k = T_k(V) = \underbrace{V \otimes \cdots \otimes V}_{k \text{ copies}}.$$

Then, as a vector space,

$$T = \oplus\{T_k \colon k \in \mathbb{N}\}.$$

There is an obvious isomorphism $\theta_{rs}\colon T_r \otimes T_s \to T_{r+s}$, determined by

$$\theta_{rs}\colon (u_1 \otimes \cdots \otimes u_r) \otimes (v_1 \otimes \cdots \otimes v_s) \mapsto u_1 \otimes \cdots \otimes u_r \otimes v_1 \cdots \otimes v_s.$$

If $x \in T_r$ and $y \in T_s$ we define a product $xy \in T_{r+s}$ via

$$xy = \theta_{rs}(x \otimes y).$$

The product is extended by linearity to all of T, and it is a routine matter to verify that then T is an (associative) F-algebra (with 1). We may view $T_0 = F$ as a subalgebra of T (so $1_T = 1_F$), and we may view $T_1 = V$ as a subspace of T. In fact T is a graded algebra (graded by $\mathbb{N} = \{0, 1, 2, \dots\}$), with T_k the subspace of homogeneous elements of degree k – the important fact is that $T_r T_s \subseteq T_{r+s}$ for all r, s.

The tensor algebra $T(V)$ is determined by the following universal mapping property. If A is an associative F-algebra and $\varphi\colon V \to A$ is a linear transformation, then (viewing V as a subspace of T) there is a unique extension of φ to an algebra homomorphism $\tilde{\varphi}\colon T \to A$.

Now let V be a quadratic space, and let $I = I(Q)$ be the ideal in $T = T(V)$ generated by all elements $v \otimes v - Q(v)$, $v \in V$, and let C be the quotient algebra T/I. When it is helpful we may write C as $C(V)$, or $C(Q)$, or even $C(V, Q)$.

Theorem 8.6. *The algebra $C = T/I$ constructed above is a Clifford algebra for the quadratic space V.*

Proof. Let $\eta\colon T \to T/I$ be the canonical quotient map, and let $\varepsilon = \eta|_V$, i.e. $\varepsilon v = v + I$, all $v \in V$. Since V generates T as an algebra εV generates C as an algebra. Note that C is compatible with Q, via ε, by the definition of the ideal I. If A is another algebra, compatible with Q via $\varphi\colon V \to A$, then φ extends uniquely to an algebra homomorphism $\tilde{\varphi}\colon T \to A$ by the universal mapping property characterization of T.

Note that $\tilde{\varphi}(v \otimes v - Q(v)) = (\varphi v)^2 - Q(v)1_A = 0$, all $v \in V$, since A is compatible with Q, and so $I \subseteq \operatorname{Ker}\tilde{\varphi}$. Thus $\tilde{\varphi}$ naturally induces a homomorphism $\widehat{\varphi}\colon C = T/I \to A$ for which $\widehat{\varphi}(v + I) = \tilde{\varphi}(v) = \varphi(v)$, i.e. $\widehat{\varphi}\varepsilon = \varphi$.

The picture is as follows.

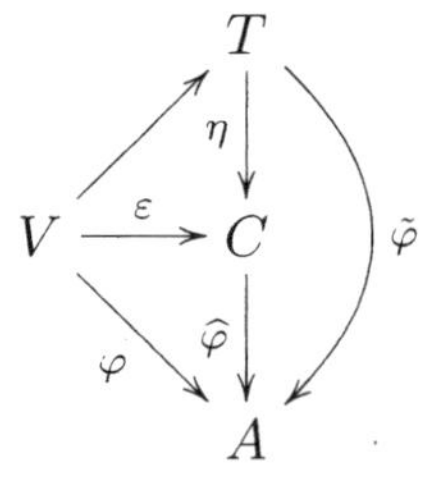

The homomorphism $\widehat{\varphi}$ is unique since $\varepsilon(V)$ generates C as an algebra. □

Proposition 8.7. *The linear transformation $\varepsilon\colon V \to C$ is 1–1.*

Proof. Note that $\operatorname{Ker}\varepsilon = V \cap \operatorname{Ker}\eta = V \cap I$. Since $V = T_1$ has no elements of degree greater than 1, whereas I is generated by elements of degree 2, it is clear that $V \cap I = 0$. □

By virtue of Proposition 8.7 we may (and shall) identify V with εV as a subspace of the Clifford algebra C. This means in particular that we may write simply uv for the product in C of u, $v \in V$ rather than the cumbersome $(u+I)(v+I) = u \otimes v + I$.

It may be instructive to rewrite the results of Proposition 8.1, Corollary 8.2, and Propositions 8.3 – 8.4 as applied to $A = C$ with the identification just made. We have:

8.1 $B(u,v) = \frac{1}{2}(uv + vu)$; hence

8.2 $u \perp v$ if and only if $uv = -vu$;

8.3 if $0 \neq v \in V$ then v is a unit in C if and only if v is anisotropic in V; and

8.4 if u is nonzero and anisotropic in V then $\sigma_u w = -uwu^{-1}$, all $w \in V$.

Exercise

Suppose that V is a quadratic space over F, that K is an extension field of F, and that $V^K = K \otimes_F V$, a K-quadratic space as in the exercise on page 36. Show that $C(V^K) \cong C(V)^K = K \otimes_F C(V)$, where $C(V)^K$ is a K-algebra in the usual fashion.

Proposition 8.8. *If $\{v_1, \dots, v_n\}$ is an orthogonal basis for V then*

$$\{v_1^{e_1} v_2^{e_2} \cdots v_n^{e_n} \colon \text{ each } e_i = 0 \text{ or } 1\}$$

is a spanning set for $C = C(V)$ as a vector space. Thus $\dim_F C \leq 2^n$.

Proof. Since V generates C as an algebra each $x \in C$ can be written as a linear combination of products of the basis vectors v_i. But since $v_i^2 = Q(v_i) \in F$, and $v_i v_j = -v_j v_i$ if $i \neq j$, each summand of such a linear combination can be rewritten as a scalar multiple of some $v_1^{e_1} v_2^{e_2} \cdots v_n^{e_n}$, with each e_i being 0 or 1. Thus the set of all elements of that form is a spanning set (over F) for C. □

Let us next investigate the effect of the grading of T on C. Write $T_+ = \oplus\{T_k \colon k \text{ even}\}$ and $T_- = \oplus\{T_k \colon k \text{ odd}\}$, so $T = T_+ \oplus T_-$. Set

$$C_0 = \eta(T_+) = (T_+ + I)/I \cong T_+/(T_+ \cap I)$$

and

$$C_1 = \eta(T_-) = (T_- + I)/I \cong T_-/(T_- \cap I),$$

so $C = C_0 + C_1$.

Proposition 8.9. $I = (I \cap T_+) \oplus (I \cap T_-)$ *(as vector spaces).*

Proof. The set of all elements of the form $\sum_i r_i(v_i \otimes v_i - Q(v_i))s_i$, where r_i, $s_i \in T$ and $v_i \in V$, is obviously an ideal of T contained in I and containing a set of generators for I, so it is I. If $x = \sum_{i=1}^k r_i(v_i \otimes v_i - Q(v_i))s_i \in I$ then each r_i and each s_i can be split via $T = T_+ \oplus T_-$, and it follows that $x = x_+ + x_-$, with $x_+ \in I \cap T_+$ and $x_- \in I \cap T_-$. Since

$$(I \cap T_+) \cap (I \cap T_-) \subseteq T_+ \cap T_- = 0$$

we conclude that $I = (I \cap T_+) \oplus (I \cap T_-)$. □

Proposition 8.10. *The Clifford algebra C splits as $C = C_0 \oplus C_1$.*

Proof. We need only show that $C_0 \cap C_1 = 0$. Set $D_0 = T_+/(T_+ \cap I)$ and $D_1 = T_-/(T_- \cap I)$, so $D_i \cong C_i$, $i = 0, 1$, as noted above. Define $\theta\colon T \to D_0 \oplus D_1$ (external direct sum) as follows. If $x \in T$ write $x = x_+ + x_-$, $x_+ \in T_+$, $x_- \in T_-$, then set $\theta x = (x_+ + D_0, x_- + D_1)$. Clearly θ is a homomorphism from T onto $D_0 \oplus D_1$. If $x \in I$ then $x = x_+ + x_-$, with $x_+ \in I \cap T_+$, $x_- \in I \cap T_-$ by Proposition 8.9, so $x \in \operatorname{Ker}\theta$, i.e. $I \subseteq \operatorname{Ker}\theta$. Thus θ naturally induces a surjective homomorphism $\widehat{\theta}\colon C = T/I \to D_0 \oplus D_1$. If $x = x_+ + x_-$ and $x + I \in \operatorname{Ker}\widehat{\theta}$ then $x_+ \in I \cap T_+$ and $x_- \in I \cap T_-$, and therefore $x \in I$ by Proposition 8.9, so $\widehat{\theta}$ is 1–1. Thus $\dim C = \dim D_0 + \dim D_1 = \dim C_0 + \dim C_1$. Since $C = C_0 + C_1$ and $\dim C$ is finite it follows that $C_0 \cap C_1 = 0$, and hence $C = C_0 \oplus C_1$. □

The decomposition $C = C_0 \oplus C_1$ of the Clifford algebra determines a $\mathbb{Z}_2$*-grading* of C. Note that $F \subseteq C_0$ and $V \subseteq C_1$. If r, $s \in \{0, 1\}$ then $C_r C_s = C_{r+s}$, the sum $r + s$ being taken mod 2. Thus C_0 is a subalgebra of C (the *even subalgebra*); the elements of C_0 are called *even homogeneous* of *degree* 0, and the elements of C_1 are called *odd homogeneous* of *degree* 1.

Exercise

Suppose that V is a quadratic space of dimension 1, let $\{x\}$ be a basis, and say that $Q(x) = a \in F^\star$. Show that $T(V)$ is isomorphic with the polynomial algebra $F[x]$, and that I corresponds to the principal ideal $\langle x^2 - a\rangle$ in $F[x]$. If $a \notin F^{\times 2}$ show that $C \cong F(\sqrt{a})$, a field. If $a = b^2$, $b \in F$, show that $C \cong F[x]/\langle x - b\rangle \oplus F[x]/\langle x + b\rangle \cong F \oplus F$. Identify C_0 and C_1 in each of the cases.

In general an F-algebra A is $\mathbb{Z}_2$*-graded* if it has subspaces A_0, A_1 such that $A = A_0 \oplus A_1$, $1 \in A_0$ (hence we assume that $F \subseteq A_0$), and if $r, s \in \{0, 1\}$ then $A_r A_s \subseteq A_{r+s}$, with the sum $r + s$ taken mod 2; A_0 is called the *even* subalgebra.

If $A = A_0 \oplus A_1$ and $B = B_0 \oplus B_1$ are $\mathbb{Z}_2$-graded F-algebras define their $\mathbb{Z}_2$-graded *tensor product* $A \widehat{\otimes} B$ as follows: as an F-vector space

$$\begin{aligned} A\widehat{\otimes}B &= A \otimes B \\ &= (A_0 \oplus A_1) \otimes (B_0 \oplus B_1) \\ &= [(A_0 \otimes B_0) \oplus (A_1 \otimes B_1)] \oplus [(A_0 \otimes B_1) \oplus (A_1 \otimes B_0)]. \end{aligned}$$

The multiplication in $A\widehat{\otimes}B$ is a twisted version of the usual multiplication for tensor products of algebras; it is determined by setting

$$(x_1 \otimes y_1)(x_2 \otimes y_2) = (-1)^{(\deg y_1)(\deg x_2)} x_1 x_2 \otimes y_1 y_2$$

for homogeneous (i.e. even or odd) y_1 and x_2, and extending by linearity.

One verifies as usual, using the universal properties of $A \otimes B$, that the multiplication is well defined, and it is easily checked to be associative. The $\mathbb{Z}_2$-grading of $A\widehat{\otimes}B$ is given by

$$\begin{aligned} (A\widehat{\otimes}B)_0 &= (A_0 \otimes B_0) \oplus (A_1 \otimes B_1), \\ (A\widehat{\otimes}B)_1 &= (A_0 \otimes B_1) \oplus (A_1 \otimes B_0). \end{aligned}$$

We may (and shall) identify A with $A\widehat{\otimes}F$ and B with $F\widehat{\otimes}B$, and hence view A and B as ($\mathbb{Z}_2$-graded) subalgebras of $A\widehat{\otimes}B$.

Let us apply these ideas to Clifford algebras.

Proposition 8.11. *Suppose that V is a quadratic space, and that $V = V_1 \textcircled{\perp} V_2$. Then $C(V) = C(V_1)\widehat{\otimes}C(V_2)$*

Proof. (Sketch.) Define $\varepsilon\colon V \to C(V_1)\widehat{\otimes}C(V_2)$ as follows: write $v \in V$ as $v = v_1 + v_2$, $v_i \in V_i$, and set $\varepsilon v = v_1 \otimes 1 + 1 \otimes v_2$. Then

$$\begin{aligned} (\varepsilon v)^2 &= (v_1 \otimes 1 + 1 \otimes v_2)(v_1 \otimes 1 + 1 \otimes v_2) \\ &= v_1^2 \otimes 1 - v_1 \otimes v_2 + v_1 \otimes v_2 + 1 \otimes v_2^2 \\ &= Q(v_1) + Q(v_2) = Q(v), \end{aligned}$$

so $C(V_1)\widehat{\otimes}C(V_2)$ is compatible with Q.

If A is an algebra compatible with Q via $\varphi\colon V \to A$ set $\varphi_i = \varphi|_{V_i}$, $i = 1, 2$. Then φ_i is compatible with $Q|_{V_i}$, so φ_i extends to an algebra homomorphism $\widehat{\varphi}_i\colon C(V_i) \to A$. Define $\varphi^\star\colon C(V_1) \times C(V_2) \to A$ via $\varphi^\star(x, y) = \widehat{\varphi}_1(x)\widehat{\varphi}_2(y)$. One checks that $\varphi^\star$ is balanced, so there is a linear transformation $\widehat{\varphi}\colon C(V_1)\widehat{\otimes}C(V_2) \to A$ such that $\widehat{\varphi}(x \otimes y) = \widehat{\varphi}_1(x)\widehat{\varphi}_2(y)$. One also checks, using the definition of multiplication in $C(V_1)\widehat{\otimes}C(V_2)$, that $\widehat{\varphi}$ is an algebra homomorphism. If $v = v_1 + v_2 \in V$ then

$$\begin{aligned} \widehat{\varphi}\varepsilon(v) &= \widehat{\varphi}(v_1 \otimes 1 + 1 \otimes v_2) \\ &= \widehat{\varphi}_1(v_1 \otimes 1) + \widehat{\varphi}_2(1 \otimes v_2) \\ &= \varphi(v_1) + \varphi(v_2) = \varphi(v), \end{aligned}$$

so $\widehat{\varphi}\varepsilon = \varphi$.

Finally, $\widehat{\varphi}$ is unique since εV includes all $v_1 \otimes 1$ and all $1 \otimes v_2$, $v_i \in V_i$, and those elements generate $C(V_1)\widehat{\otimes}C(V_2)$ as an algebra. □

Theorem 8.12. *If V is a quadratic space of dimension n then $C(V)$ is of dimension 2^n. If $\{v_1, v_2, \dots, v_n\}$ is an orthogonal basis for V then*

$$\{v_1^{e_1} v_2^{e_2} \cdots v_n^{e_n} : \text{each } e_i = 0 \text{ or } 1\}$$

is a basis for $C(V)$.

Proof. Induction on n. If $n = 1$ the theorem holds by the exercise on page 70, so suppose that $n > 1$ and that the result holds for spaces of dimension less than n. Suppose that W is a nondegenerate subspace of V, with $1 \leq \dim W < n$. Then $V = W \oplus W^{\perp}$, and if $\dim W = k$ then $\dim C(W) = 2^k$ and $\dim C(W^{\perp}) = 2^{n-k}$ by the induction hypothesis. Thus

$$\dim C(V) = \dim(C(W)\widehat{\otimes}C(W^{\perp})) = 2^k \cdot 2^{n-k} = 2^n$$

by Proposition 8.11. Since $\{v_1^{e_1} v_2^{e_2} \cdots v_n^{e_n} : \text{each } e_i = 0 \text{ or } 1\}$ is a spanning set (Proposition 8.8) it is immediate that it is a basis since it has the correct cardinality. □

Corollary 8.13. $\dim C_0(V) = \dim C_1(V) = 2^{n-1}$.

Proof. A basis element $v_1^{e_1} v_2^{e_2} \cdots v_n^{e_n}$ is in C_0 if $\sum_i e_i$ is even, and is in C_1 if $\sum_i e_i$ is odd. □

Exercise

Suppose that V is a (quadratic) hyperbolic plane, with hyperbolic pair (u, v). Set

$$\overline{u} = \begin{bmatrix} 0 & 0 \\ 1 & 0 \end{bmatrix} \quad \text{and} \quad \overline{v} = \begin{bmatrix} 0 & 1 \\ 0 & 0 \end{bmatrix}$$

in the matrix algebra $M_2(F)$. Show that $M_2(F)$ is compatible with Q via $\varphi\colon u \mapsto \overline{u}$, $v \mapsto \overline{v}$, and that $\widehat{\varphi}\colon C(V) \to M_2(F)$ is an isomorphism. (*Hint*: $\widehat{\varphi}$ is surjective.)

Proposition 8.14. *Suppose that $x \in C_0(V)$ and $xv = vx$ for all $v \in V$. Then $x \in F$.*

Proof. We may write x as a linear combination of basis vectors of the form $v_1^{e_1} v_2^{e_2} \cdots v_n^{e_n}$, each $e_i = 0$ or 1, and $\sum_i e_i$ even. Say such a specific basis vector has coefficient $a \in F$, so $x = a v_1^{e_1} v_2^{e_2} \cdots v_n^{e_n} + \cdots$. If $v_1^{e_1} v_2^{e_2} \cdots v_n^{e_n} \neq 1$ then $e_j = 1$ for some j. Thus

$$\begin{aligned} xv_j &= (a v_1^{e_1} v_2^{e_2} \cdots v_n^{e_n} + \cdots) v_j \\ &= v_j x = v_j(a v_1^{e_1} v_2^{e_2} \cdots v_n^{e_n} + \cdots) \\ &= (-a v_1^{e_1} v_2^{e_2} \cdots v_n^{e_n} \pm \cdots) v_j \end{aligned}$$

since v_j anticommutes with the odd number of $v_i \neq v_j$ with $e_i = 1$. But v_j is a unit in C, so

$$av_1^{e_1}v_2^{e_2}\cdots v_n^{e_n} + \cdots = -av_1^{e_1}v_2^{e_2}\cdots v_n^{e_n} \pm \cdots$$

and hence $a = 0$. Thus $v_1^0 v_2^0 \cdots v_n^0 = 1$ is the only basis vector that survives, and $x \in F$. □

Proposition 8.15. *If V is a quadratic space there is a unique anti-automorphism $\alpha\colon C(V) \to C(V)$ with $\alpha|_V = 1_V$.*

Proof. Let C^{op} denote the *opposite algebra* of C. Thus C and C^{op} are equal as sets and as vector spaces, but the multiplication in C^{op} is given by $x \circ y = yx$, yx being the product of y and x as usual in C. If A is any algebra compatible with Q then clearly A^{op} is also compatible with Q. Given any such A, with $\varphi\colon V \to A$, we have

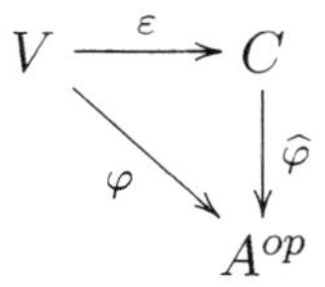

commutative for a unique homomorphism $\widehat{\varphi}$. But then also $\widehat{\varphi}$ is a unique homomorphism from C^{op} to A for which the diagram

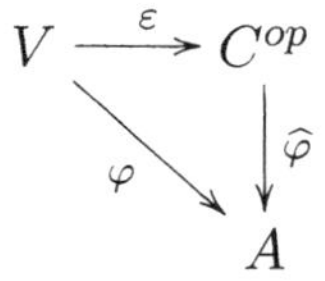

is commutative, so C^{op} is also a Clifford algebra for V. By definition there is then a unique isomorphism $\alpha\colon C \to C^{op}$ such that $\alpha|_V = 1_V$, which means that $\alpha\colon C \to C$ is an anti-automorphism. □

Observe that Proposition 8.15 is valid at characteristic 2 as well – the proof did not require our standing assumption that F have odd characteristic.

Chapter 9

Orthogonal Groups (char $F \neq 2$), II

We continue to assume that V is a quadratic space of dimension $n \geq 2$ over a field F of characteristic different from 2.

Proposition 9.1. *Suppose that $u_1, u_2, \ldots, u_k$ are nonzero anisotropic vectors in V such that $\sigma_{u_1}\sigma_{u_2}\cdots\sigma_{u_k} = 1_V$. Then $\prod_{i=1}^k Q(u_i) \in F^{\times 2}$.*

Proof. Note that k is even, since 1_V is a rotation. For any $v \in V$ we have

$$\begin{aligned} 1_V v &= v = \sigma_{u_1}\sigma_{u_2}\cdots\sigma_{u_k} v \\ &= (u_1\cdots u_k)v(u_1\cdots u_k)^{-1} \end{aligned}$$

by Proposition 8.4. But then $u_1 \cdots u_k = a \in F^\star$ by Proposition 8.14.

Now apply the anti-automorphism α of Proposition 8.15. Note first that

$$\begin{aligned} u_1\cdots u_k u_k \cdots u_1 &= Q(u_k)u_1\cdots u_{k-1}u_{k-1}\cdots u_1 \\ &= Q(u_{k-1})Q(u_k)u_1\cdots u_{k-2}u_{k-2}\cdots u_1 = \cdots . \end{aligned}$$

Continuing in that fashion we see that

$$\begin{aligned} \prod_{i=1}^k Q(u_i) &= u_1\cdots u_k u_k\cdots u_1 \\ &= u_1\cdots u_k \alpha(u_k)\cdots\alpha(u_1) \\ &= u_1\cdots u_k \alpha(u_1\cdots u_k) \\ &= a\cdot\alpha(a\cdot 1) \\ &= a^2 \in F^{\times 2}, \end{aligned}$$

since α fixes elements of F. □

Proposition 9.1 is the basis for an important definition. Take $\sigma \in \mathrm{O}(V)$. Then σ is a product of reflections in many ways, say e.g. that $\sigma = \sigma_{u_1} \cdots \sigma_{u_r}$ and $\sigma = \sigma_{v_1} \cdots \sigma_{v_s}$. Then

$$\sigma_{u_1} \cdots \sigma_{u_r} \sigma_{v_s} \cdots \sigma_{v_1} = 1,$$

and consequently

$$Q(u_1) \cdots Q(u_r) Q(v_s) \cdots Q(v_1) \in F^{\times 2}$$

by Proposition 9.1. It follows that $Q(u_1) \cdots Q(u_r)$ and $Q(v_1) \cdots Q(v_s)$ lie in the same coset of $F^{\times 2}$ in $F^{\star}$, and hence determine a well-defined map

$$\theta\colon \mathrm{O}(V) \to F^{\star}/F^{\times 2},$$

viz. $\theta(\sigma) = Q(u_1) \cdots Q(u_r) F^{\times 2}$.

The map θ is evidently a homomorphism; it is called the *spinor norm* on $\mathrm{O}(V)$.

Exercise

Calculate the spinor norm θ for $\mathrm{O}(V)$ if V is a hyperbolic plane (see Proposition 6.1 and its proof).

Proposition 9.2. *If V is isotropic then $\theta(\mathrm{SO}(V)) = F^{\star}/F^{\times 2}$.*

Proof. Clearly the image of $\mathrm{SO}(V)$ under θ consists of all cosets of the form $a_1 \cdots a_{2k} F^{\times 2}$ with $0 \neq a_i \in Q(V)$. Since V is nondegenerate Q is universal by Proposition 4.7. Thus θ maps $\mathrm{SO}(V)$ onto $F^{\star}/F^{\times 2}$. □

Proposition 9.3. *Suppose that u and v are anisotropic vectors in V with $Q(u) = Q(v)$. Then $\sigma_u \sigma_v \in \Omega(V)$.*

Proof. By the Witt Extension Theorem (5.2) there exists $\tau \in \mathrm{O}(V)$ with $\tau v = u$. Thus $\sigma_u \sigma_v = \sigma_{\tau v} \sigma_v = \tau \sigma_v \tau^{-1} \sigma_v^{-1} \in \Omega$. □

Corollary 9.4. *Suppose that $u_1 \ldots, u_k$ and $v_1 \ldots, v_k$ are nonzero anisotropic vectors in V such that $(Q(u_1), \ldots, Q(u_k))$ is a permutation of the k-tuple $(Q(v_1), \ldots, Q(v_k))$. Then $\sigma_{u_1} \cdots \sigma_{u_k}$ and $\sigma_{v_1} \cdots \sigma_{v_k}$ lie in the same coset of Ω in $\mathrm{O}(V)$.*

Proof. Relabel $v_1, \ldots, v_k$ to $w_1, \ldots, w_k$ so that $Q(w_i) = Q(u_i)$, all i. Since $\mathrm{O}(V)/\Omega$ is abelian we have

$$\sigma_{v_k} \cdots \sigma_{v_1} \sigma_{u_1} \cdots \sigma_{u_k} \Omega = (\sigma_{u_1} \sigma_{w_1}) \Omega \cdots (\sigma_{u_k} \sigma_{w_k}) \Omega = \Omega$$

by the Proposition, so

$$\sigma_{u_1} \cdots \sigma_{u_k} \Omega = \sigma_{v_1} \cdots \sigma_{v_k} \Omega.$$

□

The kernel of the restriction to $\mathrm{SO}(V)$ of the spinor norm θ, denoted by $\mho = \mho(V)$, is called the *reduced orthogonal group* of V (relative, of course, to the given quadratic form).

Note that any commutator τ in $\mathrm{O}(V)$ has the form

$$\begin{aligned}\tau &= (\sigma_{u_1}\dots\sigma_{u_k})(\sigma_{v_1}\dots\sigma_{v_\ell})(\sigma_{u_1}\dots\sigma_{u_k})^{-1}(\sigma_{v_1}\dots\sigma_{v_\ell})^{-1}\\ &= \sigma_{u_1}\dots\sigma_{u_k}\sigma_{v_1}\dots\sigma_{v_\ell}\sigma_{u_k}\dots\sigma_{u_1}\sigma_{v_\ell}\dots\sigma_{v_1},\end{aligned}$$

from which it is clear that $\theta(\tau) = F^{\times 2}$, since $F^\star/F^{\times 2}$ is abelian. Thus

$$\Omega \triangleleft \mho \triangleleft \mathrm{SO}(V) \triangleleft \mathrm{O}(V).$$

Proposition 9.5. *If* $\dim V = 2$ *then* $\mho = \Omega$.

Proof. Take $\sigma \in \mho(V) \leq \mathrm{SO}(V)$ and write $\sigma = \sigma_u\sigma_v$. Since $\sigma \in \mho$ we have $\theta(\sigma) = 1$, i.e. $Q(u)Q(v) \in F^{\times 2}$. Thus $Q(u) = a^2Q(v) = Q(av)$ for some $a \in F^{\times 2}$, and so $\sigma = \sigma_u\sigma_v = \sigma_u\sigma_{av} \in \Omega$ by Proposition 9.3. $\square$

Exercise

If $\dim V = 2$ show that $\Omega = \{\sigma^2 \colon \sigma \in \mathrm{SO}(V)\}$.

Proposition 9.6. *Suppose that* W *is a nondegenerate subspace of* V. *If* $Q(W) = Q(V)$ *and* $\Omega(W) = \mho(W)$ *then* $\Omega(V) = \mho(V)$.

Proof. Take $\tau \in \mho(V)$ and write $\tau = \sigma_{v_1}\cdots\sigma_{v_k}$. Choose $w_1, \dots, w_k \in W$ with $Q(w_i) = Q(v_i)$, $1 \leq i \leq k$, and set $\sigma = \sigma_{w_1}\cdots\sigma_{w_k}$. Then $\theta(\sigma) = \theta(\tau)$, so $\sigma \in \mho(V)$ by Corollary 9.4. Since $V = W \bigoplus W^\perp$ and each $w_i \in W$ we have $\sigma = \alpha \bigoplus 1_{W^\perp}$, $\alpha \in \mathrm{O}(W)$. But $\theta(\sigma) = \theta(\alpha)$, so $\alpha \in \mho(W) = \Omega(W)$, and hence $\sigma \in \Omega(V)$. Thus $\tau \in \sigma\Omega(V) = \Omega(V)$. $\square$

Theorem 9.7. *Suppose that* V *is isotropic. Then*

$$\mho(V) = \Omega(V), \text{ and } \mathrm{SO}(V)/\Omega \cong F^\star/F^{\times 2}.$$

Proof. Let W be a hyperbolic plane in V. Then Q and $Q|_W$ are both universal, by Proposition 4.7, i.e. $Q(V) = Q(W) = F$. Since $\mho(W) = \Omega(W)$ by Proposition 9.5 we have $\mho(V) = \Omega(V)$ by Proposition 9.6. Since Q is universal θ maps $\mathrm{SO}(V)$ onto $F^\star/F^{\times 2}$, and hence $\mathrm{SO}(V)/\operatorname{Ker}\theta = \mathrm{SO}(V)/\Omega \cong F^\star/F^{\times 2}$. $\square$

Theorem 9.8. *Suppose that* V *is a quadratic space with discriminant* $d \cdot F^{\times 2}$ *and spinor norm* θ. *Then* $\theta(-1_V) = d \cdot F^{\times 2}$, *and in particular* $-1_V \in \mho$ *if and only if* n *is even and* $\operatorname{discr}(V) = 1$.

Proof. Let $\{v_1, \dots, v_n\}$ be an orthogonal basis for V and set $\sigma = \sigma_{v_1} \cdots \sigma_{v_n}$. Note that $\sigma\, v_i = -v_i$ for each i and hence $\sigma = -1_V$. Thus

$$\theta(-1_V) = \prod_1^n Q(v_i) F^{\times 2} = dF^{\times 2}.$$

The second statement follows since $-1_V \in \mathrm{SO}(V)$ if and only if n is even. □

Corollary 9.9. *If V is isotropic then $\Omega \cap Z(\mathrm{O}(V)) = \pm 1$ if and only if n is even and* $\mathrm{discr}(V) = 1F^{\times 2}$.

Remarks

1. Let us sketch an alternative approach to $\theta|_{\mathrm{SO}(V)}$. Define the *Clifford group* $\Gamma = \Gamma(V)$ to be the normalizer in the group $U(C)$ of units in C of the subspace V of C, i.e.

$$\Gamma = \{x \in U(C) \colon xvx^{-1} \in V, \text{ all } v \in V\}.$$

Define the *even Clifford group* to be $\Gamma_0 = \Gamma \cap C_0$. If $x \in V$ is nonzero and anisotropic, and $v \in V$, then $xvx^{-1} = -\sigma_x v \in V$ by Proposition 8.3, so $\mathrm{O}(V) \leq \Gamma$. It follows easily that $\mathrm{SO}(V) \leq \Gamma_0$.

2. If $x \in \Gamma$ and $v \in V$ define $\chi_x(v) = xvx^{-1} \in V$. Then

$$Q(\chi_x v) = Q(xvx^{-1}) = (xvx^{-1})^2 = xv^2x^{-1} = xQ(v)x^{-1} = Q(v),$$

so $\chi_x \in \mathrm{O}(V)$. In fact χ is a homomorphism from Γ into $\mathrm{O}(V)$, called the *vector representation* of Γ. It can be shown that χ maps Γ_0 *onto* $\mathrm{SO}(V)$, with $\mathrm{Ker}(\chi|_{\Gamma_0}) = F^\star$.

3. If α is the unique anti-automorphism of C with $\alpha|_V = 1_V$ discussed in Proposition 8.15, then we may define a homomorphism $N \colon \Gamma_0 \to F^\star$ via $N(x) = \alpha(x)x$. The kernel of N is called the *spin group* $\mathrm{Spin}(V)$. Then $\chi(\mathrm{Spin}(V)) = \mho(V)$, the reduced orthogonal group. If $\sigma \in \mathrm{SO}(V)$ then there exists $x \in \Gamma_0$ with $\chi(x) = \sigma$, and then $\theta(\sigma) = N(x)F^{\times 2}$ gives the spinor norm of σ.

   See section 4.8 of Jacobson's *Basic Algebra II* ([**45**]) for further details.

4. Clifford algebras were first defined by Clifford [**14**] in 1878, then appeared in the discussion of "spin representations" of orthogonal groups by E. Cartan in 1913. They were rediscovered by Pauli and Dirac in 1927-28 in the study of electron spin, hence the terminology of spinor group and spinor norm (see also [**8**]). C. Chevalley [**12**] brought the theory to the generality needed here. It has since been extended to quadratic spaces over a variety of types of rings (see e.g. [**32**]).

Recall from Corollary 4.10 that if F is a finite field (char $F \neq 2$) then there are exactly two inequivalent nondegenerate quadratic forms, one with discriminant 1 (mod $F^{\times 2}$), the other with discriminant $\neq 1$ (mod $F^{\times 2}$). It will be convenient to represent the two forms by very specific matrices. Take $d \in F^\star \setminus F^{\times 2}$.

Suppose first that $n = \dim V$ is even. By Note II following Theorem 4.2 and by Proposition 4.7 we may choose orthogonal bases relative to which the forms are represented by matrices

1. $\widehat{B} = \mathrm{diag}(1, -1, 1, -1, \ldots, 1, -1)$,
2. $\widehat{B} = \mathrm{diag}(1, -1, 1, -1, \ldots, 1, -d)$.

If $n = 2k$ then the discriminants are, respectively, $(-1)^k$ and $(-1)^k d$ mod $F^{\times 2}$. If k is even these are respectively square and nonsquare; if k is odd and -1 is a square then $-d$ is a nonsquare; if -1 is a nonsquare then $-d$ is a square.

Suppose next that $n = 2k + 1$ is odd. Then we may choose orthogonal bases such that either

3. $\widehat{B} = \mathrm{diag}(1, -1, 1, -1, \ldots, 1, -1, -1)$, or
4. $\widehat{B} = \mathrm{diag}(1, -1, 1, -1, \ldots, 1, -1, -d)$.

The respective discriminants are $(-1)^{k+1}$ and $(-1)^{k+1} d$ mod $F^{\times 2}$, which are square or nonsquare in some order as in the even case above.

Recall that if $a \in F^\star$ then *scaling* the quadratic form by a means replacing Q by Q^a, where $Q^a(v) = aQ(v)$, all $v \in V$. The scaled form may or may not be equivalent with the original form, but it clearly determines precisely the same orthogonal group on V. If we replace Q by Q^d in case 4 we obtain $\widehat{B} = \mathrm{diag}(d, -d, \ldots, d, -d, -d^2)$, with discriminant $(-1)^{k+1} d^{2k+2} \equiv (-1)^{k+1}$ mod $F^{\times 2}$. Thus we may assume, for purposes of calculating $|P\Omega(V)|$, that $\widehat{B}$ has one of the diagonal forms 1, 2, or 3 above.

We may assume then that vectors of V are represented as column vectors $(a_1, b_1, a_2, b_2, \ldots)^t \in F^n$, and that Q takes one of the three forms

1. $Q(v) = a_1^2 - b_1^2 + \cdots + a_k^2 - b_k^2$,
2. $Q(v) = a_1^2 - b_1^2 + \cdots + a_k^2 - d b_k^2$, or
3. $Q(v) = a_1^2 - b_1^2 + \cdots + a_k^2 - b_k^2 - a_{k+1}^2$.

Thus the quadric cone (see page 54) $\mathcal{C}$ consists of all $v \in V$ whose components provide solutions to

1. $x_1^2 - y_1^2 + \cdots + x_k^2 - y_k^2 = 0$,
2. $x_1^2 - y_1^2 + \cdots + x_k^2 - d y_k^2 = 0$, or
3. $x_1^2 - y_1^2 + \cdots + x_k^2 - y_k^2 - x_{k+1}^2 = 0$.

The calculation of $|\,\mathrm{O}(V)|$ will depend on the number of solutions of related equations.

Write $\mathrm{O}(n,q)$, as usual, for the orthogonal groups corresponding to Q of types 1 and 3, and write $\mathrm{O}_d(n,q)$ for Q of type 2. If Q is of type 1 or 3 and $b \in F$ let $S(n,b)$ denote the number of solutions $v \in V$ to

1. $x_1^2 - y_1^2 + \cdots + x_k^2 - y_k^2 = b$ or
3. $x_1^2 - y_1^2 + \cdots + x_k^2 - y_k^2 - x_{k+1}^2 = b$,

where we continue to write $n = 2k$ or $n = 2k+1$. If Q is of type 2 let $S_d(n,q)$ denote the number of solutions to

2. $x_1^2 - y_1^2 + \cdots + x_k^2 - dy_k^2 = b$.

Proposition 9.10. *If $|F| = q$ and $b \in F$ then*

1. $S(2k,b) = \begin{cases} q^{2k-1} + q^k - q^{k-1} & \textit{if } b = 0, \\ q^{2k-1} - q^{k-1} & \textit{if } b \neq 0; \end{cases}$

2. $S_d(2k,b) = \begin{cases} q^{2k-1} - q^k + q^{k-1} & \textit{if } b = 0, \\ q^{2k-1} + q^{k-1} & \textit{if } b \neq 0; \end{cases}$

3. $S(2k+1,b) = \begin{cases} q^{2k} & \textit{if } b = 0, \\ q^{2k} - q^k & \textit{if } -b \notin F^{\times 2}, \textit{ and} \\ q^{2k} + q^k & \textit{if } -b \in F^{\times 2}. \end{cases}$

Proof. Use induction on k. If $k = 0$ (i.e. $n = 1$) then equation 3 becomes $x_1^2 - y_1^2 - x_1^2 = -y_1^2 = b$, and clearly we have $S(1,0) = 1 = q^{2\cdot 0}$, $S(1,b) = 0 = q^{2\cdot 0} - q^0$ if $-b \notin F^{\times 2}$, and $S(1,b) = 2 = q^{2\cdot 0} + q^0$ if $-b \in F^{\times 2}$. It will help in the induction to establish the case $k = 1$ ($n = 2$) as well. When $n = 2$ then equation 1 is $x_1^2 - y_1^2 = (x_1 - y_1)(x_1 + y_1) = b$. If $b = 0$ this is $x_1 = \pm y_1$, with $2q - 1$ solutions; if $b \neq 0$ it has the $q - 1$ solutions

$$x_1 = \frac{1}{2}\left(\frac{b}{c} + c\right), \quad y_1 = \frac{1}{2}\left(\frac{b}{c} - c\right), \quad c \in F^\star$$

(why?). Thus equation 1 holds for $k = 1$.

When $k = 1$ equation 2 is $x_1^2 - dy_1^2 = b$. If $b = 0$ the only solution is $x_1 = y_1 = 0$. If $b \neq 0$ set $K = F(\sqrt{d})$, a degree 2 extension. Over K the equation has the $q^2 - 1$ solutions

$$x_1 = \frac{1}{2}\left(\frac{b}{c} + c\right), \quad y_1 = \frac{1}{2\sqrt{d}}\left(\frac{b}{c} - c\right), \quad c \in K^\star.$$

If $N = N_{K/F}$ note that $N(x_1 + y_1\sqrt{d}) = x_1^2 - dy_1^2$, so the set of solutions in F consists of those pairs $(x_1, y_1) \in F \times F$ such that $x_1 + y_1\sqrt{d} \in N^{-1}(b)$, a coset of the kernel of $N\colon K^\star \to F^\star$. Since $|\,\mathrm{Ker}\ N| = q + 1$ there are $q + 1$ solutions and equation 2 holds for $k = 1$.

Now the induction step for equation 1. Write $b = a + c$, with q choices for a. Solving equation 1 is equivalent with solving both

$$x_1^2 - y_1^2 + \cdots + x_{k-1}^2 - y_{k-1}^2 = a \quad \text{and} \quad x_k^2 - y_k^2 = c = b - a.$$

If $b = 0$ then $a = c = 0$ gives

$$S(2k-2,0)S(2,0) = (q^{2k-3} + q^{k-1} - q^{k-2})(2q-1)$$

solutions, by induction, and each of the $q-1$ cases $a \neq 0$, $c = -a$ gives

$$S(2k-2,a)S(2,c) = (q^{2k-3} - q^{k-2})(q-1).$$

Thus

$$\begin{aligned} S(2k,0) &= (q^{2k-3} + q^{k-1} - q^{k-2})(2q-1) + (q-1)^2(q^{2k-3} - q^{k-2}) \\ &= q^{2k-1} + q^k - q^{k-1}, \end{aligned}$$

as needed.

If $b \neq 0$ there are three cases: $a = 0$ and $b = c \neq 0$; $a = b \neq 0$ and $c = 0$; and $a \neq 0$, $a \neq b$, $c = b - a \neq 0$, with $q - 2$ choices for a. Inductive arguments as above yield

$$\begin{aligned} S(2k,b) &= S(2k-2,0)S(2,b) + S(2k-2,b)S(2,0) \\ &\qquad + (q-2)S(2k-2,a)S(2,c) \\ &= (q^{2k-3} + q^{k-1} - q^{k-2})(q-1) + (q^{2k-3} - q^{k-2})(2q-1) \\ &\qquad + (q-2)(q^{2k-3} - q^{k-2})(q-1) \\ &= q^{2k-1} - q^{k-1}, \end{aligned}$$

and equation 1 is proved.

The rest of the proof is similar. In fact, the proof of equation 3 is easier, since the equation can be rewritten as

$$x_1^2 - y_1^2 + \cdots + x_k^2 - y_k^2 = b + x_{k+1}^2$$

and hence reduced to q equations of type 1. □

We now have the technical wherewithal to calculate orders of finite orthogonal groups at odd characteristic.

Theorem 9.11. *Suppose* char $F \neq 2$. *Then*

1. $|\,\mathrm{O}(2k,q)| = 2q^{k(k-1)}(q^k - 1)\prod_1^{k-1}(q^{2i} - 1)$,
2. $|\,\mathrm{O}_d(2k,q)| = 2q^{k(k-1)}(q^k + 1)\prod_1^{k-1}(q^{2i} - 1)$, *and*
3. $|\,\mathrm{O}(2k+1,q)| = 2q^{k^2}\prod_1^{k}(q^{2i} - 1)$.

Proof. We use induction on n, where, as above, $n = 2k + 1$ or $2k$.

If $n = 1$ ($k = 0$) then $\mathrm{O}(V) = \{\pm 1\}$ of order 2 (by convention the empty product is equal to 1). Assume then that $n > 1$ and assume the order formulas correct for dimensions $< n$. Take $n = 2k$ and consider $\mathrm{O}(2k, q)$. Choose $v \in V$ with $Q(v) = 1$. The $\mathrm{O}(V)$-orbit of v in V is $\{w \in V : Q(w) = 1\}$ by Witt's Extension Theorem (5.2), and

$$|\operatorname{Orb}(v)| = S(2k, 1) = q^{2k-1} - q^{k-1}$$

by Proposition 9.10. Since $V = \langle v \rangle \oplus \langle v \rangle^{\perp}$ we have

$$\operatorname{Stab}_{\mathrm{O}(V)}(v) \cong \mathrm{O}(\langle v \rangle^{\perp}) = \mathrm{O}(2k-1, q),$$

of order $2q^{(k-1)^2} \prod_1^{k-1}(q^{2i} - 1)$, and therefore

$$\begin{aligned} |\mathrm{O}(2k, q)| &= (q^{2k-1} - q^{k-1}) \cdot 2q^{(k-1)^2} \prod_1^{k-1}(q^{2i} - 1) \\ &= 2q^{k-1}(q^k - 1)q^{(k-1)^2} \prod_1^{k-1}(q^{2i} - 1) \\ &= 2q^{k(k-1)}(q^k - 1) \prod_1^{k-1}(q^{2i} - 1), \end{aligned}$$

and 1 is proved. The proofs of 2 and 3 are almost identical to the proof of 1, and they are omitted. □

Note that in all cases $|\mathrm{SO}(V)| = \frac{1}{2}|\mathrm{O}(V)|$.

Corollary 9.12. 1. $|\Omega(2k+1, q)| = \frac{1}{2}q^{k^2} \prod_1^k (q^{21} - 1)$;

2. $|\mathrm{P}\Omega(2k, q)| = \frac{1}{r}q^{k(k-1)}(q^k - 1) \prod_1^{k-1}(q^{2i} - 1)$, $r = (4, q^k - 1)$; *and*

3. $|\mathrm{P}\Omega_d(2k, q)| = \frac{1}{r_d}q^{k(k-1)}(q^k + 1) \prod_1^{k-1}(q^{2i} - 1)$, $r_d = (4, q^k + 1)$.

Proof. Since $[\mathrm{SO}(V) : \Omega] = [F^\star : F^{\times 2}] = 2$ (Theorem 9.7) we must divide $|\mathrm{SO}(V)|$ by 2 in all cases, and by an extra factor of 2 in just those cases when $-1 \in \Omega$. Since $-1_V \notin \mathrm{SO}(V)$ if n is odd, 1 is established. In case 2 we have $\operatorname{discr}(V) = (-1)^k$, so $-1_V \in \Omega$ if and only if either k is even or else k is odd and $-1 \in F^{\times 2}$, by Theorem 9.8. Note that -1 is the unique element of order 2 in $F^\star$, and $|F^{\times 2}| = (q-1)/2$, so $-1 \in F^{\times 2}$ if and only if $2 \,\big|\, \frac{q-1}{2}$, i.e. $4 \,\big|\, q - 1$, or $q \equiv 1 \pmod 4$. We may combine the two requirements as follows – note that $4 \,\big|\, q^k - 1$ if k is even or if k is odd and $q \equiv 1 \pmod 4$ but if k is odd and $q \equiv -1 \pmod 4$ then $q^k - 1 \equiv -2 \pmod 4$, so $4 \nmid q^k - 1$. Thus $r = (4, q^k - 1) = 4$ if $-1_V \in \Omega$, $r = 2$ if $-1_V \notin \Omega$, and 2 is established. In case 3 $\operatorname{discr}(V) = (-1)^k d$ which is a nonsquare if k is even or if k is odd and $-1 \in F^{\times 2}$; it is a square if and only if k is odd and $-1 \notin F^{\times 2}$. One checks as above that $r_d = (q^k + 1, 4) = 4$ if $-1_V \in \Omega_d$, $r_d = 2$ if $-1_V \notin \Omega_d$, and 3 is established. □

Recall that $P\Omega(3,3) = \Omega(3,3)$ was left as an exception in Theorem 6.31. Since $|\Omega(3,3)| = \frac{1}{2} \cdot 3 \cdot (3^2 - 1) = 12$ it is of course not simple. Since $S(3,0) = |\mathcal{C}| = 3^2$ by Proposition 9.10 we have $|\mathbb{P}\mathcal{C}| = (3^2 - 1)/2 = 4$, and it follows easily that $\Omega(3,3) \cong \mathrm{Alt}_4$.

Recall from Theorem 3.12 that $|\,\mathrm{PSp}(2k,q)| = \frac{1}{2}q^{k^2}\prod_1^k(q^{2i} - 1)$ if q is odd. Thus $|\,\mathrm{PSp}(2k,q)| = |\,P\Omega(2k+1,q)|$ for all k and q. It can be shown (see e.g. [**36**]) that $\mathrm{PSp}(4,q) \cong P\Omega(5,q)$ for all q, but $\mathrm{PSp}(2k,q) \not\cong P\Omega(2k+1,q)$ if $k \geq 3$. These groups, together with $\mathrm{PSL}(3,4)$ and $\mathrm{PSL}(4,2)$, are the only examples of finite simple groups that are not isomorphic but are of the same order (see [**1**] and [**2**]).

Some further facts: $P\Omega(3,q) \cong \mathrm{PSL}(2,q)$. The excluded case, in which $\dim V = n = 4$, Witt index $m = 2$, satisfies

$$P\Omega(V) \cong \mathrm{PSL}(2,F) \times \mathrm{PSL}(2,F).$$

If $\dim V = 4$ and $m = 1$ then the discriminant D of V is a nonsquare, and if $K = F(\sqrt{D})$ then $\Omega(V) \cong \mathrm{PSL}(2,K)$.

Chapter 10

Hermitian Forms and Unitary Spaces

In this chapter we begin the study of bilinear forms defined over a degree 2 field extension, such as $\mathbb{C}$ over $\mathbb{R}$. The isometries (unitary transformations) of such forms play a role, for example, in quantum mechanics, in that quantum mechanical *states* are unitary operators (see e.g. the appendix by W. Faris in [**75**]).

Much of the material in this chapter and the next is valid over arbitrary division rings (see e.g. [**20**] and [**21**]), but we consider only vector spaces over fields.

Let us assume throughout that F is a field having an automorphism α of order 2, so $\alpha \neq 1$ but $\alpha^2 = 1$. We will usually write $\alpha(a) = \overline{a}$ for all $a \in F$, and we set $F_0 = \text{Fix}(\alpha) = \{a \in F \colon \overline{a} = a\}$, so $[F \colon F_0] = 2$ (why?). The prototypical example, of course, is $F = \mathbb{C}$, with α being complex conjugation, and $F_0 = \mathbb{R}$.

If V is a vector space of dimension n over F then a *sesquilinear form* on V, relative to the automorphism α, is a function $B \colon V \times V \to F$ satisfying

1. $B(u + v, w) = B(u, w) + B(v, w)$,
2. $B(u, v + w) = B(u, v) + B(u, w)$, and
3. $B(au, v) = aB(u, v) = B(u, \overline{a}v)$

for all $u, v, w \in V$ and all $a \in F$. If in addition the form satisfies

4. $B(u, v) = \overline{B(v, u)}$

for all $u, v \in V$, then B is called a *Hermitian bilinear form*.

Although more general sesquilinear forms will appear in the next chapter, we will assume in this chapter that B is Hermitian.

As for bilinear forms we say that u is *orthogonal* to v in V, and write $u \perp v$, if $B(u,v) = 0$. If $u \perp v$ then $v \perp u$, by property 4, and the theory of Hermitian forms closely parallels the theory of symmetric bilinear forms. We omit proofs of results for Hermitian forms if they are easy variations of the proofs of corresponding results for symmetric bilinear forms.

If $\{v_1, v_2, \dots, v_n\}$ is a basis for V, $u = \sum_i a_i v_i$, $v = \sum_i b_i v_i$, and $\widehat{B} = [B(v_i, v_j)]$, then

$$B(u,v) = \sum_{i,j} a_i B(v_i, v_j)\overline{b_j} = \mathbf{a}^t \widehat{B}\, \overline{\mathbf{b}},$$

where $\mathbf{a} = (a_1, \dots, a_n)^t$ and $\overline{\mathbf{b}} = (\overline{b_1}, \dots, \overline{b_n})^t$ in F^n. Since $B(u,v) = \overline{B(v,u)}$ for all u, v we see that $\widehat{B} = \overline{\widehat{B}^t}$, and we say that $\widehat{B}$ is a *Hermitian* matrix. Conversely a Hermitian matrix $\widehat{B}$ determines a Hermitian form relative to a basis $\{v_i\}$ for V by the formula above.

By analogy with the definition of $F^{\times 2}$ define $F^{1+\alpha}$ to be $\{a\bar{a}\colon a \in F^\star\} \subseteq F_0^\star$, motivated notationally by $a^{1+\alpha} = a \cdot a^\alpha = a\bar{a}$. Clearly $F^{1+\alpha}$ is a subgroup of $F^\star$.

If we change basis from $\{v_i\}$ to $\{w_i\}$, with $w_j = \sum_i d_{ij} v_i$, then check that

$$\widehat{B}_{\{w_i\}} = D^t \widehat{B}_{\{v_i\}} \overline{D},$$

so $\det \widehat{B}_{\{w_i\}} = (\det D \cdot \overline{\det D}) \cdot \det \widehat{B}_{v_i}$.

Define the *discriminant* of B to be

$$\operatorname{discr}(B) = \begin{cases} 0 & \text{if } \det \widehat{B} = 0, \\ (\det \widehat{B}) F^{1+\alpha} \in F^\star / F^{1+\alpha} & \text{otherwise.} \end{cases}$$

Note that $\operatorname{discr}(B)$ is independent of the choice of basis. Say that B is *nondegenerate* if $\operatorname{discr}(B) \neq 0$.

For any subset $S \subseteq V$ define

$$S^\perp = \{v \in V\colon B(u,v) = B(v,u) = 0, \text{ all } u \in S\}.$$

Define the *radical* of V to be $\operatorname{rad}(V) = V^\perp$; more generally, if W is a subspace of V define $\operatorname{rad}(W)$ to be $W \cap W^\perp$. Say that V (or a subspace W) is *nondegenerate* if $\operatorname{rad} V = 0$ ($\operatorname{rad} W = 0$), or equivalently (for V) if $\det \widehat{B} \neq 0$. If W is nondegenerate then $V = W \oplus W^\perp$.

A vector space V with a nondegenerate Hermitian form B is called a *unitary space*.

If (V_1, B_1) and (V_2, B_2) are unitary spaces then an isomorphism σ from V_1 to V_2 is an *isometry* if $B_2(\sigma u, \sigma v) = B_1(u, v)$ for all $u, v \in V_1$; V_1 and V_2 are *isometric* if there is an isometry from V_1 to V_2. Further, B_1 and B_2 are called *equivalent* if there is an isometry from V_1 to V_2.

As for symmetric bilinear forms we define the (Hermitian) *quadratic form* Q corresponding to a Hermitian form B via $Q(v) = B(v, v)$, all $v \in V$. Note that $Q(av) = a\overline{a}Q(v)$ if $a \in F$. Also $Q(v) = B(v, v) = \overline{B(v, v)} \in F_0$ for all $v \in V$.

It is not clear in general that B can be recovered from Q as in the symmetric case. It can be, for example, if $F = \mathbb{C}$ and $F_0 = \mathbb{R}$, for then

$$B(u, v) = \frac{1}{4}[Q(u + v) - Q(u - v) + iQ(u + iv) - iQ(u - iv)]$$

(verify).

Notions related to isotropy carry over essentially unchanged to unitary spaces.

Proposition 10.1. *If B is a nonzero Hermitian form on V then $Q(v) \neq 0$ for some $v \in V$.*

Proof. Suppose $Q(v) = 0$, all v. Choose $u, v \in V$ with $B(u, v) = 1$. Then for any $a \in F$ we have

$$0 = Q(u + av) = B(u, av) + B(av, u) = a + \overline{a}.$$

Take $a = 1$ to see that $2 = 0$, i.e. $\operatorname{char} F = 2$. But then $\overline{a} = -a = a$ for all $a \in F$, so the automorphism α of F does not have order 2, a contradiction. □

Theorem 10.2. *A unitary space V has an orthogonal basis $\{v_1, \dots, v_n\}$, with $Q(v_i) \in F_0^\star$, say $Q(v_i) = a_i$, so $\widehat{B} = \operatorname{diag}(a_1, \dots, a_n)$.*

Proof. Apply Proposition 10.1 to choose $v_1 \in V$ with $Q(v_1) = a_1 \in F_0^\star$. Then $W_1 = \langle v_1 \rangle$ is nondegenerate, and $V = W_1 \oplus W_1^\perp$. Apply induction to $W_1^\perp$. □

Theorem 10.3. *If F is finite and V is a unitary space then V has an orthogonal basis $\{v_i\}$ for which $\widehat{B} = I$.*

Proof. Since $[F\colon F_0] = 2$ we must have $F_0 = \mathbb{F}_q$ and $F = \mathbb{F}_{q^2}$ for some prime power q, the involution is $\alpha\colon a \mapsto a^q$, $a \in F$, and the norm map $N = N_{F/F_0}$ is given by $N(a) = aa^q = a\overline{a}$. Since F is finite N maps $F^\star$ onto $F_0^\star$. If an orthogonal basis is chosen as in the proof of Theorem 10.2, with $Q(v_i) = a_i \in F_0^\star$, all i, choose $b_i \in F^\star$ such that $N(b_i) = a_i^{-1}$; then

$$Q(b_i v_i) = b_i \overline{b_i} Q(v_i) = N(b_i) a_i = a_i^{-1} a_i = 1$$

for all i. □

Corollary 10.4. *Any two unitary spaces of the same dimension over a finite field F are equivalent.*

Proposition 10.5. *If F is a finite field and V is a unitary space over F with* $\dim V \geq 2$ *then V is isotropic.*

Proof. Choose an orthogonal basis $\{v_i\}$ with $Q(v_i) = 1$ for all i. Choose $a \in F^\star$ with $N_{F/F_0}(a) = -1$, and set $v = v_1 + av_2$. Then $v \neq 0$ and $Q(v) = 1 + a\overline{a} = 1 - 1 = 0$. □

As usual a pair (u, v) of vectors from a unitary space V is called a *hyperbolic pair* if $Q(u) = Q(v) = 0$ and $B(u, v) = 1$. The subspace $\langle u, v\rangle$ is called a (unitary) *hyperbolic plane*.

Proposition 10.6. *The trace map* $\operatorname{tr}_{F/F_0}$ *maps F onto F_0.*

Proof. Write $T\colon F \to F_0$ for $\operatorname{tr}_{F/F_0}$, then T is an F_0-linear transformation of F so it is either onto or identically 0. If $T = 0$ then $T(a) = a + \overline{a} = 0$, all $a \in F$, and in particular $T(1) = 1 + 1 = 0$, and char $F = 2$. But then $\overline{a} = -a = a$, all a, and the involution α is trivial, a contradiction. □

Corollary 10.7. *If V is a unitary space and $v \in V$ then there exists $a \in F$ with $Q(v) = a + \overline{a}$.*

Proof. Since $Q(v) \in F_0$ there is some $a \in F$ with $Q(v) = \operatorname{tr}_{F/F_0}(a) = a + \overline{a}$ by the Proposition. □

Proposition 10.8. *If u is a nonzero isotropic vector in a unitary space V then there exists $v \in V$ such that (u, v) is a hyperbolic pair.*

Proof. Since V is nonsingular there exists $w \in V$ such that $B(u, w) = 1$. For each $a \in F$ set $v_a = au + w$. Then $Q(v_a) = a + \overline{a} + Q(w)$, and by Corollary 10.7 there exists $b \in F$ with $Q(v_b) = 0$. With $v = v_b$ we have $Q(v) = Q(bu + w) = 0$ and $B(u, v) = 1$. □

Exercise

Suppose that V is unitary and isotropic, and dim $V \geq 2$. If $a \in F_0$ show that $Q(v) = a$ for some $v \in V$.

Proposition 10.9. *If V is a unitary space then*

$$V = H_1 \perp \cdots \perp H_k \perp W,$$

where each H_i is a hyperbolic plane, $1 \leq i \leq k$, and W is anisotropic.

Proof. If V is anisotropic then $V = W$. If V is isotropic choose $u_1 \neq 0$ isotropic in V, then use Proposition 10.8 to choose v_1 so that (u_1, v_1) is a hyperbolic pair, and set $H_1 = \langle u_1, v_1 \rangle$. If we set $V_1 = H_1^\perp$ then $V = H_1 \oplus V_1$, and the result follows by induction on dimension, applied to V_1. □

Corollary 10.10. *If F is a finite field and V is a unitary space over F then*

$$V = H_1 \oplus \cdots \oplus H_k \oplus W,$$

where each H_i is a hyperbolic plane, $1 \leq i \leq k$, W is anisotropic, and $\dim W = 0$ *or* 1.

Proof. Proposition 10.5. □

Proposition 10.11. *Let V be any finite dimensional vector space over F, and suppose that V_1, V_2 are subspaces with* $\dim V_1 = \dim V_2$. *Then V_1 and V_2 have a common complement, i.e. there is a subspace W such that $V = V_1 \oplus W = V_2 \oplus W$.*

Proof. Induction on the codimension $k = \dim V - \dim V_i$. If $k = 0$ then $W = 0$ will do, so assume that $k > 0$ and assume the result for lower codimensions. Since $V_1 \cup V_2 \neq V$ (why?) we may choose $w \in V \backslash (V_1 \cup V_2)$, and set $U_i = V_i \oplus \langle w \rangle$, $i = 1, 2$. Then each U_i has codimension $k - 1$, and by induction $V = U_1 \oplus W_1 = U_2 \oplus W_1$ for some subspace W_1. Set $W = \langle w \rangle \oplus W_1$, then $V = V_1 \oplus W = V_2 \oplus W$. □

We prove next the Witt Extension Theorem for unitary spaces; the proof can easily be modified (and will be in chapter 12) for a quadratic space V with char $F = 2$. The sections of the proof marked by (*), (**), and (***) will be referred to in Chapter 12. Be forewarned that the proof, while not particularly deep or complicated, is rather long and drawn out, with several cases to be considered.

Theorem 10.12 (Witt's Extension Theorem). *Suppose that V is a unitary space, V_1 and V_2 are subspaces, and $\sigma\colon V_1 \to V_2$ is an isometry. Then there is an isometry $\tau\colon V \to V$ such that $\tau_{|V_1} = \sigma$.*

Proof. Set $U = \mathrm{Ker}(\sigma - 1_{V_1}) = \{v \in V_1 \colon \sigma v = v\}$, and assume for the first part of the proof that U is a hyperplane in V_1, i.e. that $\dim U = \dim V_1 - 1$. Set

$$W = \mathrm{Im}(\sigma - 1_{V_1}) = \{\sigma v - v \colon v \in V_1\},$$

so $\dim W = 1$ and thus there exists $v_0 \in V_1$ such that $W = \langle \sigma v_0 - v_0 \rangle$.

Case 1. Suppose that $V_1 \not\subseteq W^\perp$. Note that $U \subseteq W^\perp$, since if $v \in U$ then

$$\begin{aligned} B(v, \sigma v_0 - v_0) &= B(v, \sigma v_0) - B(v, v_0) \\ &= B(v, \sigma v_0) - B(\sigma v, \sigma v_0) \\ &= B(v, \sigma v_0) - B(v, \sigma v_0) = 0. \end{aligned}$$

Since U is a hyperplane in V_1 it follows that $U = V_1 \cap W^\perp$. If we choose $v \in V_1 \setminus W^\perp$ then

$$\begin{aligned} B(\sigma v_0, \sigma v - v) &= B(\sigma v_0, \sigma v) - B(\sigma v_0, v) \\ &= B(v_0, v) - B(\sigma v_0, v) \\ &= B(v_0 - \sigma v_0) \neq 0, \end{aligned}$$

and so $V_2 = \sigma V_1 \not\subseteq W^\perp$.

Since $U = \sigma U \subseteq \sigma V_1 = V_2$ we have likewise that $U = V_2 \cap W^\perp$. Choose a subspace X so that $W^\perp = U \oplus X \subseteq V_1 + X$. Then

$$V_1 \cap X = X \cap V_2 \cap W^\perp = X \cap U = 0,$$

and similarly $V_2 \cap X = 0$. Since $V_i \not\subseteq W^\perp$ we have $W^\perp \subsetneqq V_i \oplus X$, $i = 1, 2$, and consequently $V = V_1 \oplus X = V_2 \oplus X$. Before we define τ observe that if $v_1 \in V_1$ and $x \in X$ then $\sigma v_1 - v_1 \in W \subseteq X^\perp$ (since $X \subseteq W^\perp$), and therefore $B(\sigma v_1 - v_1, x) = 0$, or $B(\sigma v_1, x) = B(v_1, x)$. Thus also $B(x, \sigma v_1) = B(x, v_1)$. Now for $v \in V$ write $v = v_1 + x$, $v_1 \in V_1$ and $x \in X$, and set

$$\tau v = \tau(v_1 + x) = \sigma v_1 + x.$$

(*) Then

$$\begin{aligned} B(\tau(v_1 + x), \tau(v_1' + x')) &= B(\sigma v_1 + x, \sigma v_1' + x') \\ &= B(\sigma v_1, \sigma v_1') + B(\sigma v_1, x') + B(x, \sigma v_1') + B(x, x') \\ &= B(v_1, v_1') + B(v_1, x') + B(x, v_1') + B(x, x') \\ &= B(v_1 + x, v_1' + x'), \end{aligned}$$

and τ is an isometry. Case 1 is finished.

Case 2. Suppose that $V_1 \subseteq W^\perp$, and therefore $U \subseteq W^\perp$. If $v \in V_1$ then

$$B(\sigma v, \sigma v_0 - v_0) = B(v, v_0) - B(\sigma v, v_0) = B(v - \sigma v, v_0) = 0$$

since $v - \sigma v \in W$ and $v_0 \in V_1 \subseteq W^\perp$, and so also $V_2 = \sigma V_1 \subseteq W^\perp$. (**) Since $v_0 \in V_1$ we have

$$\begin{aligned} 0 &= B(\sigma v_0, \sigma v_0 - v_0) - B(v_0, \sigma v_0 - v_0) \\ &= B(\sigma v_0 - v_0, \sigma v_0 - v_0), \end{aligned}$$

and therefore W is a totally isotropic line, and $W \subseteq W^\perp$. By Proposition 10.11 there exists a subspace $X \subseteq W^\perp$ such that $W^\perp = V_1 \oplus X = V_2 \oplus X$.

Exactly as in Case 1 there is an isometry τ_1 of $W^\perp$ with $\tau_{1|V_1} = \sigma$ and $\tau_{1|X} = 1_X$ – in fact

$$\tau_1(W^\perp) = \tau_1(V_1 \oplus X) = \tau_1 V_1 \oplus X = V_2 \oplus X = W^\perp,$$

and $\tau_1|_{W\oplus X} = 1_{W\oplus X}$. We now need only extend τ_1 to all of V – the net effect is that we have reduced Case 2 to:

Case $2'$. $V_1 = W^\perp$ is a hyperplane in V, and $V_2 = \sigma V_1 = V_1$. For this case fix $v \in V \setminus V_1$, and define $f \in V_1^*$ via $f(v_1) = B(\sigma^{-1}v_1, v)$. Extend f to V (e.g. via $f(v) = 0$), so $f \in V^*$, and then there exists $u \in V$ such that $f(v_1) = B(v_1, u)$, all $v_1 \in V_1$, i.e.

$$f(v_1) = B(\sigma^{-1}v_1, v) = B(v_1, u),$$

and therefore $B(v_1, v) = B(\sigma v_1, u)$ for all $v_1 \in V$ (replace v by $\sigma v_1 \in V_1$). Next we choose u more carefully, so that also $Q(u) = Q(v)$. Then set $w_0 = \sigma v_0 - v_0$, so that $\langle w_0 \rangle = W$. Since $W = V_1^\perp$ we may (and shall) replace u by $u + aw_0$ (still called u), $a \in F$, and maintain the equality $B(v_1, v) = B(\sigma v_1, u)$.

Note that $u \notin V_1 = W^\perp$. For suppose that $u \in w^\perp$ – then

$$\begin{aligned} B(v_1, u) &= B(\sigma v_1, u) + B(v_1 - \sigma v_1, u) \\ &= B(\sigma v_1, u) = B(v_1, v), \end{aligned}$$

all $v_1 \in V_1$, and therefore $v - u \in V_1^\perp = W^{\perp\perp} = W$, and therefore $v \in u + W \subseteq V_1$, a contradiction. The conclusion: $B(w_0, u) \neq 0$. (***) Thus

$$Q(u + aw_0) = Q(u) + B(u, aw_0) + B(aw_0, u).$$

Apply Corollary 10.7 to choose $b \in F$ such that $b + \bar{b} = Q(v) - Q(u)$ and set $a = b/B(w_0, u)$. Then

$$\begin{aligned} Q(u + aw_0) &= Q(u) + \frac{B(u, bw_0)}{B(w_0, u)} + \frac{B(bw_0, u)}{B(w_0, u} \\ &= Q(u) + \frac{\bar{b}B(u, w_0)}{\overline{B(w_0, u)}} + \frac{bB(w_0, u)}{B(w_0, u)} \\ &= Q(u) + \bar{b} + b = Q(v). \end{aligned}$$

Replace u by $u + aw_0$ (but continue to call it u).

We now have $V = V_1 \oplus \langle v \rangle = V_1 \oplus \langle u \rangle$, and $Q(v) = Q(u)$. Extend σ to τ by setting $\tau v = u$, and verify (easily) that τ is an isometry. This finishes the first part of the proof.

The second part is shorter. We use induction on $k = \dim V_1$, with $n = \dim V$. If $k = 1$ we return to the first part of the proof, which has been done. Suppose, then, that $k > 1$, and let W be a hyperplane in V_1. By induction $\sigma_{|W}$ extends to an isometry $\sigma_1 \colon V \to V$. If $\sigma_{1|V_1} = \sigma$ we're finished. If not set $\tau_1 = \sigma_1^{-1}\sigma$, and note that $W = \{v \in V_1 \colon \tau_1 v = v\}$. Again we are back in the first part of the proof (with a possibly different V_2), and we may extend τ_1 to an isometry τ_2 of V with $\tau_{2|V_1} = \tau_1$. Set $\tau = \sigma_1\tau_2$, an isometry of V. Then $\tau_{|V_1} = \sigma_1\tau_1 = \sigma_1\sigma_1\sigma = \sigma$, and the proof is complete. □

Corollary 10.13. *Any two maximal totally isotropic subspaces of a unitary space V are isometric. The common dimension of all of them is called the* Witt index *of V, and is denoted $m = m(V)$.*

Corollary 10.14 (Witt's Cancellation Theorem). *If V_1 and V_2 are isometric subspaces of a unitary space V then $V_1^\perp$ and $V_2^\perp$ are also isometric.*

Remark

If $V = H_1 \bigoplus \cdots \bigoplus H_k \bigoplus W$, as in Proposition 10.9, then $m(V) = k$. Thus if F is finite we have $m = \lfloor n/2 \rfloor$ by the corollary to Proposition 10.9, and in general $m \leq \lfloor n/2 \rfloor$.

Proposition 10.15. *If V is an isotropic unitary space then V has a basis of isotropic vectors.*

Proof. Choose $v_1 \neq 0$ in V with $Q(v_1) = 0$, and choose $w_2, \ldots, w_n \in V \setminus v_1^\perp$ so that $\{v_1, w_2, \ldots, w_n\}$ is a basis for V. For each i, $2 \leq i \leq n$, we may choose (by Proposition 10.8) $a_i \in F$ and set $v_i = a_i v_1 + w_i$ such that (v_1, v_i) is a hyperbolic pair. Each v_i is isotropic, and $w_i = v_i - a_i v_1 \in \langle v_1, \ldots, v_n \rangle$, so $\{v_i\}$ spans V, and is therefore a basis. □

Proposition 10.16. *If V is an isotropic unitary space with* $\dim V = n \geq 3$ *then every line in V is the intersection of two hyperbolic planes.*

Proof. Suppose first that $\langle v_1 \rangle$ is an isotropic line. By Proposition 10.9 (and its proof) there is a basis $\{v_1, v_2, \ldots, v_n\}$ of isotropic vectors such that $H_i = \langle v_1, v_i \rangle$ is a hyperbolic plane, $2 \leq i \leq n$ — any pair H_i, H_j with $i \neq j$ will intersect in $\langle v_1 \rangle$.

Suppose then that $L = \langle x \rangle$, with $Q(x) \neq 0$, and choose y isotropic in V. Choose $z \in V \setminus \{x^\perp \cup y^\perp \cup \langle x, y \rangle\}$ (why is this possible?). Then $P = \langle y, z \rangle$ is a hyperbolic plane and there exists an isotropic $v \in P$ such that $B(y, v) = 1$. Choose $a \in F^\star$ such that $a + \overline{a} = 0$ and set $u = y + av \in P$; then $Q(u) = a + \overline{a} = 0$. At least two of u, v, and y are not in $x^\perp$ since $x \notin P^\perp \subseteq z^\perp$. Set $Q_u = \langle x, u \rangle$, $Q_v = \langle x, v \rangle$, and $Q_y = \langle x, y \rangle$. At least two are hyperbolic planes, and any two intersect in $\langle x \rangle = L$. □

This study of unitary geometry will be seriously utilized in the next chapter, where we undertake the study of groups of unitary transformations, which play an important role in mathematics and the sciences.

Chapter 11

Unitary Groups

Assume for this chapter that V is a unitary space, with Hermitian bilinear form B and quadratic form Q.

An isometry of V is called a *unitary transformation*, and the group of all unitary transformations is denoted by $\mathrm{U}(V)$. As usual we write $\mathrm{U}(n, F)$, $\mathrm{U}(n, q)$, etc., for corresponding groups of matrices. If a basis is chosen and $\tau \in \mathrm{GL}(V)$ is represented by the matrix T then $\tau \in \mathrm{U}(V)$ if and only if $T^t \widehat{B} \overline{T} = \widehat{B}$. Thus $(\det \tau)\overline{(\det \tau)} = 1$ if $\tau \in \mathrm{U}(V)$.

The *special unitary group* is the subgroup

$$\mathrm{SU}(V) = \{\tau \in \mathrm{U}(V)\colon \det \tau = 1\},$$

so $\mathrm{SU}(V) \lhd \mathrm{U}(V)$. In fact, if we write $N(a) = a\overline{a}$, $a \in F^\star$, i.e. $N = N_{F/F_0}|_{F^\star}$, then det is a homomorphism from $\mathrm{U}(V)$ to $F^\star$ with kernel $\mathrm{SU}(V)$ and $\mathrm{Im}(\det) \subseteq \mathrm{Ker}\, N$. If $a \in \mathrm{Ker}\, N = \{b \in F\colon b\overline{b} = 1\}$ then there exists $\tau \in \mathrm{U}(V)$ with $\det \tau = a$ (choose an orthogonal basis and let τ be represented by $\mathrm{diag}(a, 1, \ldots, 1)$), so $\mathrm{U}(V)/\mathrm{SU}(V) \cong \mathrm{Ker}\, N = \{b \in F\colon b\overline{b} = 1\}$.

Suppose that τ is a transvection with fixed hyperplane W, and suppose that $\tau \in \mathrm{U}(V)$. Say $W^\perp = \langle u \rangle$. Exactly as for symplectic transvections (Chapter 3) there exist $z \in W$ and $y \in V$ such that $\tau v = v + B(v, y)z$, all $v \in V$. In particular $\tau w = w = w + B(w, y)z$, all $w \in W$, so $y \in W^\perp$, say $y = \overline{b}u$, $b \in F$, and so $\tau v = v + bB(v, u)z$. Since $\tau \in \mathrm{U}(V)$ we have

$$\begin{aligned} B(v, w) &= B(\tau v, \tau w) \\ &= B(v + bB(v, u)z, w + bB(w, u)z) \\ &= B(v, w) + bB(v, u)B(z, w) \\ &\qquad + \overline{b\, B(w, u)}B(v, z) + b\overline{b}B(v, u)\overline{B(w, u)}Q(z), \end{aligned}$$

and therefore

$$(*)\quad bB(v,u)B(z,w)+\overline{b\,B(w,u)}B(v,z)+b\bar{b}B(v,u)\overline{B(w,u)}Q(z)=0,$$

all $v, w \in V$. Take $v = z \perp u$ to see that $\overline{B(w,u)}Q(z) = 0$, all w, and therefore $Q(z) = 0$, i.e. z is isotropic, and so (*) becomes

$$(**)\qquad bB(v,u)B(z,w)+\overline{bBS(w,u)}B(v,z)=0,$$

all $v, w \in V$. Fix v so that $B(v,u) \neq 0$. Since $b \neq 0$ and $B(v,\cdot)$ is not identically 0 we may conclude that $B(v,z) \neq 0$, and that

$$B(w,u)=-[\bar{b}B(u,v)/bB(z,v)]B(w,z)=cB(w,z),$$

all $w \in V$. Thus $u - \bar{c}z \in V^{\perp} = 0$, and $u = \bar{c}z$. Set $a = b/\bar{c}$ and we have $\tau v = v + aB(v,u)u$, and then (**) becomes

$$\begin{aligned} aB(v,u)B(u,w)+\overline{aB(w,u)}B(v,u) &= B(v,u)[aB(u,w)+\bar{a}B(u,w)] \\ &= B(v,u)B(u,w)(a+\bar{a})=0, \end{aligned}$$

all $v, w \in V$, and therefore $a = -\bar{a}$, i.e. a is *antisymmetric* in F.

Conclusion: if τ is a transvection in $\mathrm{U}(V)$ then $m(V) > 0$, τ has fixed hyperplane $W = u^{\perp}$, with u isotropic, and there exists $a \neq 0$ in F, with $a = -\bar{a}$, so that $\tau v = v + aB(v,u)u$, all $v \in V$. Conversely, each such $u \in V$ and $a \in F$ determine a transvection in $\mathrm{U}(V)$, which we denote by $\tau_{u,a}$. Thus

$$\tau_{u,a}v = v + aB(v,u)u,$$

all $v \in V$, with $a = -\bar{a} \neq 0$ in F, $0 \neq u \in V$, and $Q(u) = 0$, describes all transvections in $\mathrm{U}(V)$.

Note that in fact $\tau_{u,a} \in \mathrm{SU}(V)$.

Suppose next that $u \in V$ with $Q(u) \neq 0$, and again set $W = u^{\perp}$. If $\sigma \in \mathrm{U}(V)$ and $\sigma|_W = 1_W$, then $\sigma u = au$, some $a \in F^{\star}$, and $Q(u) = Q(\sigma u) = a\bar{a}Q(U)$, so $a\bar{a} = 1$.

On the other hand, if $1 \neq a \in F$, with $a\bar{a} = 1$, and $u \in V$, with $Q(u) \neq 0$, set

$$\sigma_{u,a}v = v + (a-1)\frac{B(v,u)}{Q(u)}u,$$

all $v \in V$. A routine calculation shows that $\sigma_{u,a}|_W = 1_W$, $\sigma_{u,a}u = au$, and $\sigma_{u,a} \in \mathrm{U}(V)$.

We call $\sigma_{u,a}$ a *quasi-reflection along u* (or *through W*). Note that if char F is odd then $\sigma_{u,-1} = \sigma_u$, the usual reflection along u. Even if char $F = 2$ then nontrivial quasi-reflections along u exist for each nonisotropic vector u, for we may choose $b \in F$ with $b \neq \bar{b}$ and set $a = \bar{b}/b$. Then $1 \neq \sigma_{u,a} \in \mathrm{U}(V)$.

Exercises

1. Suppose that $u \in V$ is isotropic, $a, b \in F^\star$ with $a = -\bar{a}$, $b = -\bar{b}$, $c \in F^\star$, and $\sigma \in \mathrm{U}(V)$. Show that
   (a) $\tau_{u,a+b} = \tau_{u,a}\tau_{u,b}$, and therefore $\tau_{u,a}^{-1} = \tau_{u,-a}$;
   (b) $\tau_{cu,a} = \tau_{u,c\bar{c}a}$; and
   (c) $\sigma\tau_{u,a}\sigma^{-1} = \tau_{\sigma u,a}$.
2. Suppose that $u \neq 0$ in V is anisotropic, $a, b \in F$ with $a\bar{a} = b\bar{b} = 1$, $c \in F$, and $\tau \in \mathrm{U}(V)$. Show that
   (a) $\sigma_{u,ab} = \sigma_{u,b}\sigma_{u,a}$, and therefore $\sigma_{u,a}^{-1} = \sigma_{u,1/a}$;
   (b) $\sigma_{cu,a} = \sigma_{u,a}$; and
   (c) $\tau\sigma_{u,a}\tau^{-1} = \sigma_{\tau u,a}$.

Consider next a very useful generalization, due to G. E. Wall ([**72**], 1959), of the formulas for unitary transvections $\tau_{u,a}$ and quasi-reflections $\sigma_{u,a}$.

For any $\sigma \in \mathrm{U}(V)$ write $\widehat{\sigma} = 1 - \sigma$, and set $V_\sigma = \widehat{\sigma}V = \{v - \sigma v \colon v \in V\}$. Since $\sigma \in \mathrm{U}(V)$ an easy calculation establishes that

$$(*) \qquad B(\widehat{\sigma}x, y) + B(x, \widehat{\sigma}y) = B(\widehat{\sigma}x, \widehat{\sigma}y)$$

for all $x, y \in V$.

If $u = \widehat{\sigma}x$ and $v = \widehat{\sigma}y$ in V_σ define

$$B_\sigma(u, v) = B_\sigma(\widehat{\sigma}x, \widehat{\sigma}y) = B(x, \widehat{\sigma}y).$$

To see that B_σ is well defined suppose also that $u = \widehat{\sigma}x_1$ and $v = \widehat{\sigma}y_1$, and use $(*)$ above:

$$\begin{aligned} B(x, \widehat{\sigma}y) &= B(x, \widehat{\sigma}y_1) \\ &= B(\widehat{\sigma}x, \widehat{\sigma}y_1) - B(\widehat{\sigma}x, y_1) \\ &= B(\widehat{\sigma}x_1, \widehat{\sigma}y_1) - B(\widehat{\sigma}x_1, y_1) = B(x_1, \widehat{\sigma}y_1). \end{aligned}$$

It follows immediately that B_σ is a sesquilinear form (see page 85) on V_σ relative to the field automorphism α, $\alpha(a) = \bar{a}$.

In place of condition 4 on page 85 it is easy to verify (using $(*)$) that we have instead the condition

$$(**) \qquad B_\sigma(u, v) + \overline{B_\sigma(v, u)} = B(u, v)$$

for all $u, v \in V_\sigma$.

If $B_\sigma(u, v) = 0$ for all $u \in V_\sigma$ then $B(x, \widehat{\sigma}y) = 0$ for all $x \in V$, and thus $v = \widehat{\sigma}y = 0$, i.e. $\mathrm{rad}_R(V_\sigma) = 0$, and B_σ is nondegenerate by Proposition 2.1.

Let $\{v_1, \ldots, v_r\}$ be a basis for V_σ, set $\widehat{B_\sigma} = [B_\sigma(v_i, v_j)]$, the representing matrix as usual, and set $A = [a_{ij}] = \widehat{B_\sigma}^{-1}$. For any $x \in V$ write

$$\widehat{\sigma}(x) = \sum_1^r \lambda_i v_i,$$

$\lambda_i = \lambda_i(x) \in F$. By the definition of B_σ we have

$$B(x, v_j) = B_\sigma(\widehat{\sigma}x, v_j) = \sum_i \lambda_i B_\sigma(v_i, v_j),$$

and hence $\lambda_j = \sum_i B(x, v_i)a_{ij}$ for each j. Thus

$$x - \sigma x = \widehat{\sigma}x = \sum_{i,j} B(x, v_i)a_{ij}v_j,$$

and hence

$$\sigma x = x - \sum_{i,j} B(x, v_i)a_{ij}v_j,$$

all $x \in V$. Note also that $\sigma|_{V_\sigma^\perp}$ is the identity.

Now let us reverse the above process.

Proposition 11.1. *Suppose that V is a unitary space with Hermitian bilinear form B, that W is a subspace of V, and that C is a nondegenerate sesquilinear form on W satisfying*

$$(**) \qquad C(u, v) + \overline{C(v, u)} = B(u, v)$$

for all $u, v \in W$. Then there is a unique $\sigma \in \mathrm{U}(V)$ such that $W = V_\sigma$ and $C = B_\sigma$.

Proof. Uniqueness will follow from existence, since, relative to any given basis for W, σ will be expressed by the formula preceding the proposition. Choose a basis $\{v_1, \ldots, v_r\}$ for W and *define* $\sigma : W \to W$ by the formula

$$\sigma x = x - \sum_{i,j} B(x, v_i)a_{ij}v_j,$$

where $A = [a_{ij}] = \widehat{C}^{-1}$. Note that the condition $(**)$ can be written as

$$A^{-1} + \overline{A^t}^{-1} = [B(v_i, v_j)],$$

and consequently

$$(***) \qquad A[B(v_i, v_j)]\overline{A^t} = A + \overline{A^t}.$$

Clearly σ is a linear transformation; let us verify that $\sigma \in \mathrm{U}(V)$. If $x, y \in V$ then

$$\begin{aligned} B(\sigma x, \sigma y) &= B(x - \sum_{i,j} B(x, v_i)a_{ij}v_j, y - \sum_{k,\ell} B(y, v_k)a_{k\ell}v_\ell) \\ &= B(x,y) - \sum_{i,j} B(x, v_i)a_{ij}B(v_j, y) - \sum_{k,\ell} B(v_k, y)\overline{a_{k\ell}}B(x, v_\ell) \\ &\quad + \sum_{i,j,k,\ell} B(x, v_i)a_{ij}B(v_j, v_\ell)\overline{a_{k\ell}}B(v_k, y). \end{aligned}$$

Replace ℓ by i and k by j in the third term above, and reverse the roles of j and k in the fourth term, and $B(\sigma x, \sigma y)$ becomes

$$B(x,y) - \sum_{i,j} B(x, v_i)(a_{ij} + \overline{a_{ji}})B(v_j, y)$$

$$+\sum_{i,j} B(x, v_i)(\sum_{k,\ell} a_{ik}B(v_k, v_\ell)\overline{a_{j\ell}})B(v_j, y),$$

which is equal to $B(x, y)$ by $(***)$. □

Exercise

Complete the proof of Proposition 11.1 by showing that $V_\sigma = W$ and that $B_\sigma = C$.

Denote the transformation $\sigma \in \mathrm{U}(V)$ in Proposition 11.1 by $\sigma_{W,C}$. Thus relative to a basis $\{v_i\}$ for W we have

$$\sigma_{W,C}(x) = x - \sum_{i,j} B(x, v_i)a_{ij}v_j,$$

all $x \in V$, where $A = \widehat{B}_\sigma^{-1} = \widehat{C}^{-1}$.

Suppose for example that $\dim W = 1$, say $W = \langle u \rangle$. Then $\sigma_{W,C}(x) = x - aB(x,u)u$, where $a = C(u,u)^{-1}$, and by $(**)$ we have $a^{-1} + \overline{a^{-1}} = Q(u)$. If u is isotropic that becomes $a + \overline{a} = 0$, and $\sigma_{W,C}$ is just the transvection $\tau_{u,-a}$. If u is anisotropic then $\sigma_{W,C}$ is the quasi-reflection $\sigma_{u,c}$, with $c = 1 - aQ(u)$.

Proposition 11.2. *Given a subspace W of the unitary space V and a sesquilinear form C on W as above, $\sigma = \sigma_{W,C}$, and $\tau \in \mathrm{U}(V)$, then*

$$\begin{aligned} V_{\tau\sigma\tau^{-1}} &= \tau V_\sigma = \tau W, \quad \text{and} \\ \tau\sigma_{W,C}\tau^{-1} &= \sigma_{\tau W, C_\tau}, \end{aligned}$$

where $C_\tau(\tau u, \tau v) = C(u,v)$.

Proof. Calculate. □

Suppose now that $W = V_\sigma$ for some $\sigma \in \mathrm{U}(V)$ and let W_1 be a nondegenerate B_σ-subspace of W. Set $W_2 = \perp_L (W_1)$ in W. Then $W = W_1 \oplus W_2$ by Proposition 2.9 (slightly generalized). By Proposition 11.1 there exist $\sigma_1, \sigma_2 \in \mathrm{U}(V)$ such that $V_{\sigma_i} = W_i$ relative to the sesquilinear forms $B_\sigma|_{W_i}$, $i = 1, 2$.

Proposition 11.3. *With hypotheses as in the paragraph above $\sigma = \sigma_2\sigma_1$.*

Proof. Choose bases $u_1, \dots, u_s$ for W_1 and $v_1, \dots, v_t$ for W_2 so that $\{u_i\} \cup \{v_j\}$ is a basis for W. Relative to that basis $\widehat{B}_\sigma$ has the form $\begin{bmatrix} L & N \\ 0 & M \end{bmatrix}$, with L and M the matrices representing $B_\sigma|_{W_1}$ and $B_\sigma|_{W_2}$, respectively. The ij-entry of N is

$$B_\sigma(u_i, v_j) = B(u_i, v_j) - \overline{B_\sigma(v_j, u_i)} = B(u_i, v_j)$$

by $(**)$. Say that $L^{-1} = [\alpha_{ij}]$ and $M^{-1} = [\beta_{ij}]$, and observe that

$$\begin{bmatrix} L & N \\ 0 & M \end{bmatrix}^{-1} = \begin{bmatrix} L^{-1} & -L^{-1}NM^{-1} \\ 0 & M^{-1} \end{bmatrix}.$$

Now calculate: for any $x \in V$ we have

$$\begin{aligned} \sigma_1 x &= x - \sum_{i,j} B(x, u_i)\alpha_{ij}u_j, \\ \sigma_2 x &= x - \sum_{k,\ell} B(x, v_k)\beta_{k\ell}v_\ell, \end{aligned}$$

and so

$$\begin{aligned} \sigma_2\sigma_1 x &= \sigma_2(x - \sum_{i,j} B(x, u_i)\alpha_{ij}u_j), \\ &= x - \sum_{i,j} B(x, u_i)\alpha_{ij}u_j - \sum_{k,\ell} B(x, v_k)\beta_{k\ell}v_\ell \\ &\quad + \sum_{i,j,k,\ell} B(x, u_i)\alpha_{ij}B(u_j, v_k)\beta_{k\ell} = \sigma x, \end{aligned}$$

since $\sum_{j,k} \alpha_{ij}B(u_j, v_k)\beta_{k\ell}$ is the $i\ell$-entry of $L^{-1}NM^{-1}$. $\square$

Remark

A converse to the proposition above is true, but it won't be needed for the following results.

Proposition 11.4. *If C is a sesquilinear form on W relative to a field automorphism α of order 2 on F then there is a basis $\{w_i\}$ for W for which $\widehat{C}$ is upper triangular, i.e. $C(w_i, w_j) = 0$ if $i \geq j$.*

Proof. If $C \equiv 0$ then $\widehat{C} = 0$. If $C \not\equiv 0$ then exactly as in Proposition 10.1 we may choose w_1 with $C(w_1, w_1) \neq 0$. Set $W_1 = \perp_L(w_1)$. Then $W = \langle w_1 \rangle \oplus W_1$, and by induction there exists a basis $w_2, \dots, w_k$ for W_1 such that $C(w_i, w_j) = 0$ if $2 \leq j < i \leq k$. Since $C(w_i, w_1) = 0$ if $2 \leq i \leq k$ the proof is complete. □

Remark

Note that the first basis vector w_1 in the proof above could be chosen arbitrarily in W, provided that $C(w_1, w_1) \neq 0$.

Proposition 11.5. *If $1 \neq \sigma \in U(V)$ we may factor σ as $\sigma = \sigma_r \cdots \sigma_2 \sigma_1$, where each σ_i is either a transvection or a quasi-reflection. The vector along which σ_1 acts can be chosen to be any $u \in V$ for which $B_\sigma(u) \neq 0$.*

Proof. Choose a basis $\{w_i\}$ for $W = V_\sigma$ as in Proposition 11.4, with $w_1 = u$. By Proposition 11.3 we have $\sigma = \tau_1 \sigma_1$, with σ_1 corresponding to $\langle w_1 \rangle$ and $B_\sigma|_{\langle w_1 \rangle}$, and τ_1 corresponding to $\langle w_2, \dots, w_r \rangle$ and $B_\sigma|_{\langle w_2, \dots, w_r \rangle}$. As remarked earlier (see page 97) σ_1 is a transvection if w_1 is isotropic, otherwise it is a quasi-reflection. Induction completes the proof. □

Proposition 11.6. $Z(U(V)) = \{c \cdot 1_V \colon c \in F$ *and* $c\bar{c} = 1\} \cong \operatorname{Ker} N_{F^\star/F_0^\star}$.

Proof. Clearly $\{c \cdot 1_V \colon c \in F,\ c\bar{c} = 1\} \leq Z = Z(U(V))$. If $\tau \in Z$ then τ commutes with every unitary transvection $\tau_{u,a}$. Thus $\tau \tau_{u,a} \tau^{-1} = \tau_{\tau u, a} = \tau_{u,a}$, and therefore $\tau u = c \cdot u$, some $c \in F$, so $\tau_{\tau u, a} = \tau_{u, c\bar{c}a} = \tau_{u,a}$, and hence $c\bar{c} = 1$. Also τ commutes with every unitary quasi-reflection $\sigma_{u,a}$. Thus $\tau \sigma_{u,a} \tau^{-1} = \sigma_{\tau u, a} = \sigma_{u,a}$, and therefore $\tau u = cu$ for some $c \in F$, so $\sigma_{\tau u, a} = \sigma_{u, c\bar{c}a} = \sigma_{u,a}$, and hence again $c\bar{c} = 1$. Thus for every line $\langle u \rangle$ in V we have $\tau u = c_u \cdot u$, with $c_u \overline{c_u} = 1$. By the usual argument $c_u = c$ is independent of (nonzero) $u \in V$, so $\tau = c \cdot 1_V$, with $c\bar{c} = 1$. □

Assume for the remainder of the chapter that V is a unitary space with Witt index $m = m(V) > 0$.

Write $\mathcal{T} = \mathcal{T}(V)$ for the subgroup of $\mathrm{U}(V)$ generated by unitary transvections. By Exercise 1(c) on page 95 we have $\mathcal{T} \triangleleft \mathrm{U}(V)$; clearly $\mathcal{T} \leq \mathrm{SU}(V)$.

Denote by $\mathcal{C}$ the *Hermitian quadric cone* in V, i.e.

$$\mathcal{C} = \{v \in V \colon Q(v) = 0\},$$

and by $\mathbb{P}\mathcal{C}$ the corresponding *projective Hermitian quadric cone*,

$$\mathbb{P}\mathcal{C} = \{[v] \in \mathbb{P}_{n-1}(V) \colon Q(v) \neq 0\}.$$

Proposition 11.7. *The subgroup $\mathcal{T}$ is transitive on the projective cone $\mathbb{P}\mathcal{C}$.*

Proof. Choose $[u] \neq [v]$ in $\mathbb{P}\mathcal{C}$, and assume first that $B(u,v) \neq 0$, so in fact we may assume that $B(u,v) = 1$, i.e. that (u,v) is a hyperbolic pair. Choose $c \in F^\star$ with $c+\bar{c} = 0$, and set $x = u+cv \neq 0$; then $Q(x) = c+\bar{c} = 0$, and so $x \in \mathcal{C}$. Set $a = -1/\bar{c}$. Then

$$\begin{aligned} \tau_{x,a}(u) &= u + aB(u,x)x \\ &= u - (1/\bar{c})B(u, u+cv)(u+cv) \\ &= u - (u+cv) = -cv \in \langle v \rangle, \end{aligned}$$

and therefore $\tau_{x,a}[u] = [v]$.

Suppose next that $B(u,v) = 0$ (then $m(V) \geq 2$, $n \geq 4$). Then choose $y \in V \setminus (u^\perp \cup v^\perp)$, so $B(u,y) \neq 0$, $B(v,y) \neq 0$. Thus we may assume that $B(u,y) = B(v,y) = 1$ by scaling u and v. We may then replace y by $y + cu$ for some $c \in F$ so that $Q(y) = 0$, hence (u,y) and (v,y) are hyperbolic pairs. By the first part of the proof there exist τ_1, $\tau_2 \in \mathcal{T}$ with $\tau_1[u] = [y]$ and $\tau_2[y] = [v]$, and therefore $\tau_2\tau_1[u] = [v]$. □

Proposition 11.8. *If V is a hyperbolic plane then $\mathcal{T} = \mathcal{T}(V)$ is doubly transitive on $\mathbb{P}\mathcal{C}$.*

Proof. Let (u,v) be a hyperbolic pair in V, and take $[x] \neq [y]$ in $\mathbb{P}\mathcal{C}$. It will suffice to exhibit $\tau \in \mathcal{T}$ with $\tau[x] = [u]$ and $\tau[y] = [v]$. By Proposition 11.7 we may in fact assume that $[x] = [u]$ and exhibit $\tau \in \mathcal{T}$ such that $\tau[u] = [u]$ and $\tau[y] = [v]$. Write $y = au + bv$; then $Q(y) = a\bar{b} + \bar{a}b = 0$. Note that $b \neq 0$ since $[y] \neq [x] = [u]$. Set $c = -a/b$, note that $\bar{c} = -c$, and set $\tau = \tau_{u,c}$. Then

$$\tau y = au + bv + cbu = au + bv - ab^{-1}bu = bv,$$

so $\tau[y] = [v]$, and of course $\tau[u] = [u]$. □

Proposition 11.9. *If V is a hyperbolic plane then $\mathcal{T} = \mathcal{T}(V) = \mathrm{SU}(V)$, i.e. $\mathrm{SU}(V)$ is generated by unitary transvections.*

Proof. Let (u,v) be a hyperbolic pair. Choose $a \in F^\star$ with $\bar{a} = -a$, and choose any $b \in F_0^\star$; then define $\tau_b \in \mathcal{T}$ by

$$\tau_b = \tau_{v,-a}\tau_{u,a^{-1}(1-b^{-1})}\tau_{v,ab}\tau_{u,a^{-1}(b^{-2}-b^{-1})}.$$

An easy (matrix) calculation shows that $\tau_b u = bu$, $\tau_b v = b^{-1}v$, i.e. τ_b is represented by the matrix $\begin{bmatrix} b & 0 \\ 0 & b^{-1} \end{bmatrix}$ relative to the basis $\{u,\, v\}$.

Now take any $\sigma \in \mathrm{SU}(V)$. By Proposition 11.8 there exists $\rho \in \mathcal{T}$ with $\rho\sigma[u] = [u]$ and $\rho\sigma[v] = [v]$. Since $\mathcal{T} \leq \mathrm{SU}(V)$ we must have $\rho\sigma u = bu$ and $\rho\sigma v = b^{-1}v$ for some $b \in F^\star$, i.e. $\rho\sigma$ is represented by $\begin{bmatrix} b & 0 \\ 0 & b^{-1} \end{bmatrix}$. Then

$$1 = B(u, v) = B(\rho\sigma u, \rho\sigma v) = B(bu, b^{-1}v) = b\bar{b}^{-1},$$

so $b = \bar{b} \in F_0^\star$, and $\rho\sigma = \tau_b \in \mathcal{T}$, as above. Thus $\sigma = \rho^{-1}\tau_b \in \mathcal{T}$, and the proof is complete. □

Proposition 11.10. *If V is a hyperbolic plane then*

$$\mathrm{SU}(V) = \langle \tau_{u,a},\ \tau_{v,b} \colon a, b \in F^\star,\ \bar{a} = -a,\ \bar{b} = -b \rangle,$$

where (u, v) is a quadratic pair.

Proof. If $0 \neq w = au + bv$ then w is isotropic if and only if $Q(w) = a\bar{b} + \bar{a}b = 0$. Take any $c \in F^\star$ with $\bar{c} = -c$. If $b = 0$ then $\tau_{w,c} = \tau_{au,c} = \tau_{u,ca\bar{a}}$, so suppose that $b \neq 0$. Then it follows from $a\bar{b} + \bar{a}b = 0$ that $\overline{ab^{-1}} = -ab^{-1}$, and then

$$\tau_{u,ab^{-1}}\tau_{v,cb\bar{b}}\tau_{u,ab^{-1}}^{-1} = \tau_{v+ab^{-1}u,cb\bar{b}} = \tau_{bv+au,c} = \tau_{w,c}.$$

Thus $\mathrm{SU}(V) = \mathcal{T} \leq \langle \tau_{u,a}, \tau_{v,b} \rangle \leq \mathrm{SU}(V)$. □

Theorem 11.11. *If $(n, |F|) \neq (3, 4)$ then $\mathcal{T}$ is transitive on the set of hyperbolic planes in V.*

Proof. We may assume that $n \geq 3$. Choose a hyperbolic plane H with hyperbolic pair (u, v). If H_1 is any other hyperbolic plane, with hyperbolic pair (u_1, v_1) we wish to exhibit $\sigma \in \mathcal{T}$ with $\sigma H_1 = H$. By Proposition 11.7 there is no loss of generality in assuming that $u_1 = u$. If $H_1 = H$ we may take $\sigma = 1$, so we may assume that $\dim(H + H_1) = 3$. As a basis for $H + H_1$ take u, v, and $w = au + v - v_1$, where $a = B(v_1, v)$. Check that $w \perp u$, $w \perp v$, $B(v_1, w) = a + \bar{a}$, and $B(w, w) = -(a + \bar{a})$. Set $W = \langle u, w \rangle$ and define a sesquilinear form C on W so that relative to the basis $\{u, w\}$ we have $\widehat{C} = \begin{bmatrix} 0 & 1 \\ -1 & -\bar{a} \end{bmatrix}$. Note that then

$$\begin{aligned} \widehat{C} + \overline{\widehat{C}^t} &= \begin{bmatrix} 0 & 1 \\ -1 & -\bar{a} \end{bmatrix} + \begin{bmatrix} 0 & -1 \\ 1 & -a \end{bmatrix} \\ &= \begin{bmatrix} 0 & 0 \\ 0 & -a - \bar{a} \end{bmatrix} = \begin{bmatrix} B(u, u) & B(u, w) \\ B(w, u) & B(w, w) \end{bmatrix}. \end{aligned}$$

Take $\sigma = \sigma_{W,C}$. Then $A = \widehat{C}^{-1} = \begin{bmatrix} -\bar{a} & -1 \\ 1 & 0 \end{bmatrix}$, and so

$$\sigma x = x + B(x, u)\bar{a}u + B(x, u)w - B(x, w)u.$$

In particular $\sigma u = u$ and $\sigma v_1 = v_1 + \overline{a}au + w - (a + \overline{a})u = v$, so $\sigma H_1 = H$. It remains to be shown that $\sigma \in \mathcal{T}$.

Write $\sigma = \sigma_2\sigma_1$ as in Proposition 11.3. If w is isotropic then W is totally isotropic, so σ_1 and σ_2 must be transvections, and $\sigma \in \mathcal{T}$. Suppose then that $Q(w) = -(a + \overline{a}) \neq 0$, and assume σ_1 chosen along w. Then in fact

$$\sigma_1 = \sigma_{w,-a\overline{a}^{-1}} \text{ and } \sigma_2 = \sigma_{\overline{a}u+w,-\overline{a}a^{-1}},$$

as can be seen by the proof of Proposition 11.3, or by computing $\sigma_2\sigma_1$ directly. The idea of the rest of the proof is to find $\rho \in \mathcal{T}$ such that $\sigma_2 = \rho\sigma_1^{-1}\rho^{-1}$, and hence $\sigma = \rho(\sigma_1^{-1}\rho^{-1}\sigma_1) \in \mathcal{T}$.

If we assume first that $|F| \neq 4$ then we may choose $c \in F_0 \setminus \{0, 1\}$ and set $\beta = ((1 - c)a)/(c(a + \overline{a}))$. Choose $\alpha \in F$ such that $\alpha + \overline{\alpha} = \beta\overline{\beta}(a + \overline{a})$ (recall that the trace function maps F onto F_0). Then set $w_1 = \alpha u + v + \beta w$ and check that $H_2 = \langle u, w_1 \rangle$ is a hyperbolic plane with hyerbolic pair (u, w_1). Define a sesquilinear form C_2 on H_2 such that $\widehat{C}_2 = \begin{bmatrix} 0 & 1 - c^{-1} \\ \overline{c}^{-1} & 0 \end{bmatrix}$ relative to basis $\{u, w_1\}$, and observe that

$$\widehat{C}_2 + \overline{\widehat{C}_2^t} = \begin{bmatrix} 0 & 1 \\ 1 & 0 \end{bmatrix}.$$

Then $A_2 = \widehat{C}_2^{-1} = \begin{bmatrix} 0 & c \\ c(c-1)^{-1} & 0 \end{bmatrix}$, and we obtain $\rho \in \mathrm{U}_{H_2}(V)$, with

$$\rho x = x - B(x, u)cw_1 - B(x, w_1)\frac{c}{c - 1}u,$$

all x. In particular $\rho u = (1 - c)^{-1}u$ and $\rho w_1 = (1 - c)w_1$, so $\rho \in \mathrm{SU}_{H_2}(V)$. Thus $\rho \in \mathcal{T}$ by Proposition 11.9. Furthermore

$$\begin{aligned}
\rho w &= w - B(w, w_1)\frac{c}{c - 1}u \\
&= w - B(w, \alpha u + v + \beta w)\frac{c}{c - 1}u \\
&= w + \frac{(a + \overline{a})(1 - c)\overline{a}c}{c(a + \overline{a})(1 - c)}u = \overline{a}u + w,
\end{aligned}$$

so $\rho\sigma_1^{-1}\rho^{-1} = \rho\sigma_{w,-\overline{a}a^{-1}}\rho^{-1} = \sigma_{\overline{a}u+w,-\overline{a}a^{-1}} = \sigma_2$, completing the proof in this case.

Suppose finally that $|F| = 4$, and hence $n \geq 4$, $m(V) \geq 2$. Thus $H^\perp$ is an isotropic unitary space of dimension at least 2. Note that $a \neq \overline{a}$, or else $a + \overline{a} = 2a = 0$, and w would be isotropic. Choose $u_3 \in H^\perp$, isotropic, so that $B(w, u_3) = \overline{a}$, and set $v_3 = \overline{a}^{-1}(w - u_3)$. Check that $Q(v_3) = 0$ and $B(u_3, v_3) = 1$, so (u_3, v_3) is a hyperbolic pair for $H_3 = \langle u_3, v_3 \rangle$, and note that $w = u_3 + \overline{a}v_3$. Set $\eta = \tau_{u_3,1}\tau_{u+u_3,1} \in \mathcal{T}$, and calculate that $\eta w = \overline{a}u + w$.

Just as above we have then that $\sigma = \sigma_2\sigma_1 = \eta\sigma_1^{-1}\eta^{-1}\sigma_1 \in \mathcal{T}$, and the theorem is proved. □

Corollary 11.12. *$\mathcal{T}$ is transitive on the set of hyperbolic pairs in V.*

If $\rho \in \mathrm{SU}(V)$ then ρ is called a *hyperbolic rotation* if there is a hyperbolic plane $H \subseteq V$ such that $\rho|_{H^\perp} = 1_{H^\perp}$, i.e. $\rho = \rho|_H \oplus 1_{H^\perp}$. Note in particular that every unitary transvection is a hyperbolic rotation.

Denote by $\mathcal{R} = \mathcal{R}(V)$ the subgroup of $\mathrm{SU}(V)$ generated by hyperbolic rotations, so $\mathcal{T} \leq \mathcal{R} \leq \mathrm{SU}(V)$. If $n = 2$ then the three groups are equal by Proposition 11.9.

Proposition 11.13. *If $|F| = 4$ and $n = \dim V = 4$ then $\mathcal{R}(V) = \mathrm{SU}(V)$.*

Proof. By Corollary 10.10 we have $V = H_1 \perp H_2$, with each H_i a hyperbolic plane. Take $\sigma \in \mathrm{SU}(V)$. By Theorem 11.11 there exists $\eta \in \mathcal{T} \leq \mathcal{R}$ such that $\eta\sigma H_1 = H_1$. Since $\eta\sigma \in \mathcal{R}$ if and only if $\sigma \in \mathcal{R}$ we may assume to start that $\sigma H_1 = H_1$. Choose $x \in H_1$ nonzero and anisotropic. Then $\sigma x \in H_1$ and $Q(\sigma x) = Q(x)$. By the Witt Extension Theorem (10.12) there exists $\tau \in U(H_1)$ with $x = \tau\sigma x$; say $\det \tau = a$, so $a\bar{a} = 1$. Choose $z \in H_1$, $z \neq 0$, with $z \perp x$. Define $\gamma \in \mathrm{GL}(H_1)$ via $\gamma x = x$, $\gamma z = a^{-1}z$, and note that γ is a quasi-reflection, so $\gamma \in U(H_1)$, and that $\det \gamma = a^{-1}$. Thus $\gamma\tau \in \mathrm{SU}(H_1)$, and if we view γ and τ as extended trivially to H_2, hence in $U(V)$, we see that $\gamma\tau \in \mathcal{R}$, $\gamma\tau\sigma x = x$, and $\gamma\tau\sigma H_1 = H_1$. Since $\gamma\tau\sigma \in \mathcal{R}$ if and only if $\sigma \in \mathcal{R}$ we may assume that $\sigma x = x$ as well as $\sigma H_1 = H_1$. Choose $w \in H_1 \cap x^\perp$, $w \neq 0$, and note that $x^\perp = \langle w \rangle \perp H_2$. Since $x^\perp$ is nondegenerate w is anisotropic, and we may choose an isotropic vector $u \in x^\perp$ with $B(u, w) = 1$ (take $v_1 \in H_2$ with $Q(v_1) = -Q(w)$ and set $u = w + v_1$). Thus $H_3 = \langle u, w \rangle$ is also a hyperbolic plane. Since $\sigma x = x$, $\sigma H_1 = H_1$, $w \in H_1$, and $w \perp x$, we have $\sigma w = bw$, $b \in F^\star$, with $b\bar{b} = 1$. Say that $H_3 \cap H_2 = \langle v \rangle$, and note that $\{w, v\}$ is an orthogonal basis for H_3 (and therefore v is anisotropic). Define $\delta \in \mathrm{SU}(H_3)$ via $\delta w = b^{-1}w$, $\delta v = \bar{b}^{-1}v$, and extend to V by making δ trivial on $H_3^\perp$, so δ is a hyperbolic rotation, $\delta \in \mathcal{R}$. Then $\delta\sigma x = x$ (since $x \in H_3^\perp$) and $\delta\sigma w = b^{-1}bw = w$, and $\delta\sigma_{|H_1} = 1_{H_1}$. But then $\delta\sigma$ is a hyperbolic rotation relative to $H_2 = H_1^\perp$, so $\delta\sigma \in \mathcal{R}$, and finally $\sigma \in \mathcal{R}$. □

Proposition 11.14. *If $n \geq 2$ then, except for $n = 3$ and $|F| = 4$, $\mathrm{SU}(V)$ is generated by hyperbolic rotations, i.e. $\mathrm{SU}(V) = \mathcal{R}$.*

Proof. Induction on n. The case $n = 2$ was Proposition 11.9, so we may assume that $n > 2$, and if $|F| = 4$ we may also assume that $n > 4$ by Proposition 11.13. If P is any hyperbolic plane in V we may choose $x \in P^\perp$ that is nonzero and anisotropic, and then $x^\perp$ is isotropic and nondegenerate. Choose a hyperbolic plane $H \subseteq V$ with $x \in H$. Now take $\sigma \in \mathrm{SU}(V)$. By

Proposition 11.10 there exists $\eta \in \mathcal{T} \leq \mathcal{R}$ such that $\eta\sigma H = H$. Since we wish to show that $\eta \in \mathcal{R}$, we may assume that $\sigma H = H$. Thus $x, \sigma x \in H$, and $Q(\sigma x) = Q(x)$. By the Witt Extension Theorem (10.12) there exists $\tau \in \mathrm{U}(H)$ with $\tau x = \sigma x$. Exactly as in the proof of Proposition 11.13 we may produce $\gamma \in \mathrm{U}(V)$ and extend τ to $\mathrm{U}(V)$ so that $\gamma\tau \in \mathcal{R}$, $\gamma\tau\sigma x = x$, and $\gamma\tau H = H$, and therefore we may assume that in fact $\sigma x = x$ and $\sigma H = H$. Thus $\sigma|_{x^\perp} \in \mathrm{SU}(x^\perp)$. By induction $\sigma|_{x^\perp} \in \mathcal{R}(x^\perp)$; since $\sigma x = x$ it follows that $\sigma \in \mathcal{R}$. □

Remark

The application of Proposition 11.10 in the proof of Proposition 11.13 precluded the induction back to $n = 2$ when $|F| = 4$. The case of $\mathrm{SU}(3, 4)$ really *is* an exception.

Theorem 11.15. *If* $n \geq 2$ *but* $(n, |F|) \neq (3, 4)$ *then* $\mathrm{SU}(V) = \mathcal{T}(V)$, *i.e.* $\mathrm{SU}(V)$ *is generated by transvections.*

Proof. If ρ is a hyperbolic rotation then there is a hyperbolic plane $H \subseteq V$ such that the restriction of ρ to $H^\perp$ is $1_{H^\perp}$ and $\rho|_H \in \mathrm{SU}(H)$. But $\mathrm{SU}(H) = \mathcal{T}(H)$ by Proposition 11.9, so $\rho \in \mathcal{T}$. Thus $\mathrm{SU}(V) = \mathcal{R} \leq \mathcal{T}$, and so $\mathrm{SU}(V) = \mathcal{T}$. □

Proposition 11.16. *If* $n \geq 2$ *then the kernel of the action of* $\mathrm{U}(V)$ *on* $\mathbb{P}\mathcal{C}$ *is* $Z(\mathrm{U}(V)) = \{a1 \colon a \in F \text{ and } a\bar{a} = 1\}$.

Proof. If $n = 2$ and $\{u, v\}$ is a basis for V then

$$\mathbb{P}\mathcal{C} = \{[u], [v], [u + bv] \colon \bar{b} = -b \neq 0\}.$$

If τ is in the kernel of the action and τ is represented by $\begin{bmatrix} a & c \\ d & e \end{bmatrix}$, then $c = 0$ since $\tau[u] = [u]$, $d = 0$ since $\tau[v] = [v]$, and $e = \bar{a}a^{-1}$, since $\tau \in \mathrm{U}(V)$. As a result $a\bar{a} = 1$, since $\tau[u + bv] = [u + bv]$, and $\tau = a1$.

If $n > 2$ suppose that σ is in the kernel. Then σ fixes every isotropic line, and so σ fixes every hyperbolic plane. By Proposition 10.16 σ fixes *every* line in V, so by the usual argument $\sigma = a1$. Since $\sigma \in \mathrm{U}(V)$ we have $a\bar{a} = 1$. It is of course clear that $Z(\mathrm{U}(V))$ is contained in the kernel of the action. □

Proposition 11.17. *The center of* $\mathrm{SU}(V)$ *is*

$$\mathrm{SU}(V) \cap Z(\mathrm{U}(V)) = \{a1 \colon a \in F,\ a\bar{a} = 1, \text{ and } a^n = 1\}.$$

Proof. Clearly $\mathrm{SU}(V) \cap Z(\mathrm{U}(V)) \leq Z(\mathrm{SU}(V))$. If $n = 2$ and V has hyperbolic pair (u, v), and $\bar{a} = -a$, $\bar{b} = -b$ in $F^\star$, then $\tau_{u,a}$ and $\tau_{v,b}$ are represented by $\begin{bmatrix} 1 & a \\ 0 & 1 \end{bmatrix}$ and $\begin{bmatrix} 1 & 0 \\ b & 1 \end{bmatrix}$, respectively. An easy calculation shows that if $\sigma \in \mathrm{SU}(V)$ commutes with both $\tau_{u,a}$ and $\tau_{v,b}$ then $\sigma = c1$ for some $c \in F$, and $c^2 = c\,\overline{c} = 1$ since $\sigma \in \mathrm{SU}(V)$.

If $n \geq 3$ and $\sigma \in Z(\mathrm{SU}(V))$ then σ commutes with every transvection $\tau_{u,a}$. Thus $\sigma\tau_{u,a}\sigma^{-1} = \tau_{\sigma u,a} = \tau_{u,a}$ for every isotropic $u \in V$, and it follows easily that $\sigma\langle u \rangle = \langle \sigma u \rangle = \langle u \rangle$, i.e. σ fixes every isotropic line. Thus $\sigma \in Z(\mathrm{U}(V))$ by Proposition 11.16. □

Corollary 11.18. *The kernel of the action of* $\mathrm{SU}(V)$ *on* $\mathbb{PC}$ *is* $Z(\mathrm{SU}(V))$*; consequently* $\mathrm{PSU}(V) = \mathrm{SU}(V)/Z(\mathrm{SU}(V))$ *acts faithfully on* $\mathbb{PC}$.

We require next a few results involving very low dimensions and/or restrictions on the sizes of the fields of scalars. The results are, of course, rather special, but they are intended to fit in with, or to complement, the much more general results that will follow.

Proposition 11.19. *If V is a hyperbolic plane and $|F| > 9$ then the derived group* $\mathrm{SU}'(V)$ *is equal to* $\mathrm{SU}(V)$.

Proof. Let (u, v) be a hyperbolic pair in V and compute with matrices relative to (u, v). Note that $|F_0| \geq 4$, and also that if $b \in F_0^\star$ with $b\,\overline{b} = 1$ then $\begin{bmatrix} b & 0 \\ 0 & b^{-1} \end{bmatrix} \in \mathrm{SU}(2, F)$. Given any $c \in F^\star$ with $\overline{c} = -c$ choose $b \in F^\star \setminus \{\pm 1\}$, and set $a = c/(b^2 - 1)$. Then $\overline{a} = -a$, and $\begin{bmatrix} 1 & a \\ 0 & 1 \end{bmatrix}$ represents a transvection. Thus

$$\begin{bmatrix} b & 0 \\ 0 & b^{-1} \end{bmatrix} \begin{bmatrix} 1 & a \\ 0 & 1 \end{bmatrix} \begin{bmatrix} b & 0 \\ 0 & b^{-1} \end{bmatrix}^{-1} \begin{bmatrix} 1 & a \\ 0 & 1 \end{bmatrix}^{-1}$$

$$= \begin{bmatrix} 1 & (b^2 - 1)a \\ 0 & 1 \end{bmatrix} = \begin{bmatrix} 1 & c \\ 0 & 1 \end{bmatrix} \in \mathrm{SU}'(2, F).$$

Similarly $\begin{bmatrix} 1 & 0 \\ c & 1 \end{bmatrix} \in \mathrm{SU}'(2, F)$, and so $\mathrm{SU}(2, F) \leq \mathrm{SU}'(2, F)$ by Proposition 11.10. □

Proposition 11.20. *If $n = 3$ and $|F| = 9$ then* $\mathrm{SU}'(V) = \mathrm{SU}(V)$.

Proof. By Theorem 11.15 it will suffice to show that if u is isotropic and $0 \neq \overline{a} = -a \in F$ then $\tau_{u,a} \in \mathrm{SU}'(V)$. If $H \subseteq V$ is a hyperbolic plane with $u \in H$, say with hyperbolic pair (u, v), then $V = H \oplus H^\perp$, and $H^\perp = \langle w \rangle$, with $Q(w) \neq 0$. We may assume that $Q(w) = 1$. Calculate with matrices relative to the basis $\{u, v, w\}$. Choose $b \in F$ such that $b\overline{b} = 1$, and set $c = -b^{-2}$.

Check that $\begin{bmatrix} b & 0 & 0 \\ 0 & -b & 0 \\ 0 & 0 & c \end{bmatrix} \in \mathrm{SU}(3,9)$, and note that $\tau_{u,a}$ is represented by $\begin{bmatrix} 1 & a & 0 \\ 0 & 1 & 0 \\ 0 & 0 & 1 \end{bmatrix}$, $\tau_{u,-a/2}$ is represented by $\begin{bmatrix} 1 & -a/2 & 0 \\ 0 & 1 & 0 \\ 0 & 0 & 1 \end{bmatrix}$. Check then that the commutator

$$\begin{bmatrix} b & 0 & 0 \\ 0 & -b & 0 \\ 0 & 0 & c \end{bmatrix} \begin{bmatrix} 1 & -a/2 & 0 \\ 0 & 1 & 0 \\ 0 & 0 & 1 \end{bmatrix} \begin{bmatrix} b & 0 & 0 \\ 0 & -b & 0 \\ 0 & 0 & c \end{bmatrix}^{-1} \begin{bmatrix} 1 & -a/2 & 0 \\ 0 & 1 & 0 \\ 0 & 0 & 1 \end{bmatrix}^{-1}$$

is equal to

$$\begin{bmatrix} 1 & a & 0 \\ 0 & 1 & 0 \\ 0 & 0 & 1 \end{bmatrix},$$

and so $\tau_{u,a} \in \mathrm{SU}'(V)$. □

Proposition 11.21. $\mathrm{SU}'(4,4) = \mathrm{SU}(4,4)$.

Proof. Write $F = \{0, 1, a, a+1\}$, with $\overline{a} = a + 1 = a^2$. By Corollary 10.10 $V = H_1 \bigoplus H_2$, where each H_i is a hyperbolic plane, say with hyperbolic pair (u_i, v_i). Use $\{u_1, u_2, v_1, v_2\}$ as a basis for V, so that $\widehat{B} = \begin{bmatrix} 0 & I \\ I & 0 \end{bmatrix}$. The only nonzero antisymmetric element of F is 1, so every transvection has the form $\tau_{u,1}$ and has order 2 – we will write it simply as τ_u. By Proposition 11.7 and Theorem 11.15 it will suffice to show that there is a transvection in $\mathrm{SU}'(V)$. Check that $\tau_{u_1}\tau_{u_2}$ is represented by $\begin{bmatrix} I & I \\ 0 & I \end{bmatrix}$ and that $\tau_{u_2}\tau_{u_1+u_2}$ is represented by $\begin{bmatrix} I & S_1 \\ 0 & I \end{bmatrix}$, with $S_1 = \begin{bmatrix} 1 & 1 \\ 1 & 0 \end{bmatrix}$. Set $L = \begin{bmatrix} 1 & 1 \\ a+1 & a \end{bmatrix}$. If η is represented by $\begin{bmatrix} L & 0 \\ 0 & (\overline{L}^{-1})^t \end{bmatrix}$ check that η is in $\mathrm{SU}(V)$ and that $\eta\tau_{u_1}\tau_{u_2}\eta^{-1}$ is represented by $\begin{bmatrix} I & S_2 \\ 0 & I \end{bmatrix}$, $S_2 = LI\overline{L^t} = \begin{bmatrix} 0 & 1 \\ 1 & 0 \end{bmatrix}$. By Proposition 11.7 there exist $\sigma_1, \sigma_2 \in \mathrm{SU}(V)$ such that $\sigma_1 u_1 = u_2$ and $\sigma_2 u_2 = u_1 + u_2$. Thus

$$\begin{aligned} \eta\tau_{u_1}\tau_{u_2}\eta^{-1} &= \eta(\tau_{u_1}\tau_{\sigma u_1})\eta^{-1} \\ &= \eta(\tau_{u_1}\sigma_1\tau_{u_1}^{-1}\sigma_1^{-1})\eta^{-1} \in \mathrm{SU}'(V), \end{aligned}$$

and

$$\tau_{u_2}\tau_{u_1+u_2} = \tau_{u_2}\sigma_2\tau_{u_2}^{-1}\sigma_2^{-1} \in \mathrm{SU}'(V).$$

But then $\tau_{u_2}\tau_{u_1+u_2}\eta\tau_{u_1}\tau_{u_2}\eta^{-1}$ is in $\mathrm{SU}'(V)$, and is represented by the matrix

$$\begin{bmatrix} I & S_1+S_2 \\ 0 & I \end{bmatrix},$$

which clearly represents a transvection, since $S_1+S_2=\begin{bmatrix} 1 & 0 \\ 0 & 0 \end{bmatrix}$ (in fact it represents τ_{u_1}). □

Theorem 11.22. *Suppose that $n \geq 2$ but $(n,|F|) \neq (2,4)$, $(2,9)$, or $(3,4)$. Then $\mathrm{SU}'(V)=\mathrm{SU}(V)$.*

Proof. By Theorem 11.15 it will suffice to show that every transvection $\tau_{u,a}$ is in $\mathrm{SU}'(V)$. The isotropic vector u is contained in a hyperbolic plane H. If $|F| \neq 4$ or 9 then $\tau_{u,a}|_H$ is in $\mathrm{SU}'(H)$ by Proposition 11.19, and hence $\tau_{u,a} \in \mathrm{SU}'(V)$. If $|F|=9$ then $n \geq 3$ and there is a subspace $W_1 = H \oplus \langle w\rangle$, with $Q(w)\neq 0$, and $\tau_{u,a}|_{W_1} \in \mathrm{SU}'(W_1)$ by Proposition 11.20, hence $\tau_{u,a} \in \mathrm{SU}'(V)$. If $|F|=4$ then $n\geq 4$ and V has a subspace $W_2 = H \oplus H_1$, where H_1 is also a hyperbolic plane, and Proposition 11.21 applies. □

Proposition 11.23. *If $n \geq 3$ and $[x]$, $[y] \in \mathbb{P}\mathcal{C}$ then there exists $[w]\in\mathbb{P}\mathcal{C}$ for which $B(x,w)\neq 0$ and $B(y,w)\neq 0$.*

Proof. Suppose first that $B(x,y)\neq 0$. We may assume that $B(x,y)=1$. Choose $z\in\langle x,y\rangle^{\perp}$, nonzero and anisotropic, and set $w=x+ay+z$, where $a\neq 0$ and $a+\bar{a}=-Q(z)$. Then $Q(w)=a+\bar{a}+Q(z)=0$, so $w\in\mathbb{P}\mathcal{C}$ and $B(x,w)=\bar{a}\neq 0$, $B(y,w)=1$.

Suppose next that $B(x,y)=0$. Choose $f\in V^*$ such that $f(x)=f(y)=1$ and get $u\in V$ such that $B(x,u)=B(y,u)=1$ (Corollary 2.2). Set $w=u+bx$, with $b+\bar{b}=-Q(u)$. Then $Q(w)=Q(u)+b+\bar{b}=0$, so $[w]\in\mathbb{P}\mathcal{C}$, and $B(x,w)=B(y,w)=1$. □

Proposition 11.24. *If $(n,|F|)\neq(3,4)$ then $\mathrm{SU}(V)$ acts primitively on $\mathbb{P}\mathcal{C}$.*

Proof. This follows from Proposition 11.8 if $n=2$ so assume that $n\geq 3$ (in fact $n\geq 4$ if $|F|=4$). Suppose that $\mathcal{B}\subseteq\mathbb{P}\mathcal{C}$, with $|\mathcal{B}|\geq 1$, with $\sigma\mathcal{B}\cap\mathcal{B}$ equal either to $\mathcal{B}$ or to $\emptyset$ for each $\sigma\in\mathrm{SU}(V)$. We show first that there exist $[u]$, $[v]\in\mathcal{B}$ such that (u,v) is a hyperbolic pair. It is sufficient to find $[u]$, $[v]\in\mathcal{B}$ with $B(u,v)\neq 0$. Note that if $m(V)=1$ then $B(u,v)\neq 0$ whenever $[u]\neq[v]$ in $\mathbb{P}\mathcal{C}$, so for this part of the proof we may assume that $m\geq 2$, and hence $n\geq 4$. Take $[x]$, $[v]\in\mathcal{B}$ and suppose that $x\perp v$. Choose $w\in x^{\perp}\setminus v^{\perp}$ with $B(w,v)=1$. If $Q(w)=0$ set $u=w$; if not set $u=w+av\in x^{\perp}$, with $a+\bar{a}=-Q(w)$. Then $H=\langle u,v\rangle$ is a hyperbolic plane. By Proposition 11.7 there exists $\tau\in\mathrm{SU}_H(V)$ such that $\tau[v]=[u]$, and $\tau[x]=[x]$ since $x\in H^{\perp}$. Thus $[x]\in\mathcal{B}\cap\tau\mathcal{B}$, so $\tau\mathcal{B}=\mathcal{B}$ and therefore $[u]=\tau[v]\in\mathcal{B}$.

So, in general take $[u]$, $[v] \in \mathcal{B}$ such that (u, v) is a hyperbolic pair. Choose $[x] \in \mathbb{P}\mathcal{C}$ such that $[x] \neq [u]$. Then choose $[y] \in \mathbb{P}\mathcal{C}$ such that $B(x, y) \neq 0$ and $B(u, y) \neq 0$ (Proposition 11.23). By scaling first y, then x, we may assume that (x, y) and (u, y) are hyperbolic pairs. By Corollary 11.12 there exists $\tau \in \mathrm{SU}(V)$ with $\tau u = u$ and $\tau v = y$, and there exists $\sigma \in \mathrm{SU}(V)$ with $\sigma u = x$ and $\sigma y = y$. Then $[u] = \tau[u] \in \mathcal{B} \cap \tau\mathcal{B}$, so $\tau\mathcal{B} = \mathcal{B}$, and hence $[y] = \tau[v] \in \mathcal{B}$, so $\sigma\mathcal{B} = \mathcal{B}$, and hence $[x] = \sigma[u] \in \mathcal{B}$. Since $[x]$ was arbitrary $\mathcal{B} = \mathbb{P}\mathcal{C}$. □

Corollary 11.25. $\mathrm{PSU}(V)$ *is primitive on* $\mathbb{P}\mathcal{C}$.

For any $u \in \mathcal{C}$ set $K_u = \{\tau_{u,a} : \bar{a} = -a \in F\} \leq \mathrm{SU}(V)$, and write $K_{[u]}$ for the image of K_u in $\mathrm{PSU}(V)$. Note that $K_u \triangleleft \mathrm{Stab}_{\mathrm{SU}(V)}(u)$ (and therefore $K_{[u]} \triangleleft \mathrm{Stab}_{\mathrm{PSU}(V)}(u)$) by Exercise 1(c) on page 95, that $\langle \cup\{K_u : u \in \mathcal{C}\}\rangle = \mathrm{SU}(V)$ if $n \geq 2$ but $(n, |F|) \neq (3, 4)$ by Theorem 11.15, and that $K_{[u]}$ is abelian.

Theorem 11.26. *Suppose that* V *is a unitary space with* $\dim V = n \geq 2$ *and Witt index* $m \geq 1$. *Then* $\mathrm{PSU}(V)$ *is simple, except for* $(n, |F|) = (2, 4)$, $(2, 9)$, *or* $(3, 4)$.

Proof. Iwasawa's Theorem (0.5). □

$\mathrm{PSU}(V)$ is the twisted group of Chevalley type ${}^2A_{n-1}(F)$. See Carter [**11**], page 268.

We next make preparations for the calculation of the orders of finite unitary groups.

If F is a finite field with $|F| = q^2$, and V is a unitary space of finite dimension n over F, write i_n for $|\mathcal{C} \cup \{0\}|$, the number of isotropic (or zero) vectors in V, and write h_n for the number of hyperbolic pairs in V.

Proposition 11.27. *If* $n \geq 1$ *and* $|F| = q^2$ *then*

$$\begin{aligned} i_n &= q^{2n-1} + (-1)^n(q^n - q^{n-1}) \text{ and} \\ h_n &= (i_n - 1)q^{2n-3} = q^{2n-3}(q^n - (-1)^n)(q^{n-1} + (-1)^n). \end{aligned}$$

Proof. If $n = 1$ then clearly $i_n = 1$ and $h_n = 0$. We use induction, taking $n \geq 2$. Choose $v \in V$ with $Q(V) = 1$, so $V = \langle v \rangle \bigoplus v^\perp$. Then $x = av + w$, with $a \in F$ and $w \in v^\perp$, is in $\mathcal{C}(V)$ if *(i)* $a = 0$ and $w \in \mathcal{C}(v^\perp)$ (there are i_{n-1} such vectors), or *(ii)* $w \notin \mathcal{C}(v^\perp)$ and $a\bar{a} = -Q(w) \in F_0^\star$. In Case *(ii)* the number of w's is $q^{2(n-1)} - i_{n-1}$. For each fixed w there is a solution for a (since the norm map N is surjective), and in fact the number of solutions is $|\operatorname{Ker} N_{F^\star/F_0^\star}| = (q^2-1)/(q-1) = q+1$. Thus

$$\begin{aligned} i_n &= i_{n-1} + (q+1)(q^{2(n-1)} - i_{n-1}) \\ &= (q+1)q^{2(n-1)} - q i_{n-1}. \end{aligned}$$

By induction

$$\begin{aligned} i_n &= (q+1)q^{2(n-1)} - q[q^{2n-3} + (-1)^{n-1}(q^{n-1} - q^{n-2})] \\ &= q^{2n-1} + (-1)^n(q^n - q^{n-1}). \end{aligned}$$

In order to count hyperbolic pairs take u to be one of the $i_n - 1$ isotropic vectors in V. Choose v such that (u, v) is a hyperbolic pair, set $H = \langle u, v\rangle$, and write $V = H \perp H^\perp$. Set $v_1 = au+bv+w$, where $w \in H^\perp$. If (u, v_1) is to be a hyperbolic pair we need $B(u, v_1) = b = 1$ and $Q(v_1) = a+\bar{a}+Q(w) = 0$. For each of the $q^{2(n-2)}$ choices for $w \in H^\perp$ there are $q = |\operatorname{Ker}(\operatorname{tr}_{F/F_0})|$ solutions for $a \in F$ satisfying $a + \bar{a} + Q(a) = 0$. Thus there are $q \cdot q^{2(n-2)} = q^{2n-3}$ hyperbolic pairs with q fixed, and hence $(i_n - 1)q^{2n-3} = h_n$. □

Theorem 11.28. *If* $|F| = q^2$ *and* $\dim V = n$ *then*

$$|\operatorname{U}(V)| = q^{n(n-1)/2} \cdot \prod_{j=1}^{n} (q^j - (-1)^j).$$

Proof. If $n = 1$ then $a1 \in \operatorname{U}(1, F)$ if and only if $a\bar{a} = 1$, so $|\operatorname{U}(1, F)| = q+1$. When $n = 2$ consider first $\operatorname{SU}(V)$; take (u, v) to be a hyperbolic pair in V. If $\sigma \in \operatorname{Stab}_{\operatorname{SU}(V)}([u])$ say that $\sigma u = cu$, $\sigma v = au + bv$. Then

$$1 = B(u, v) = B(cu, au + bv) = c\bar{b},$$

so $b = 1/\bar{c}$. Thus $\det\sigma = 1 = c/\bar{c}$, and $c = \bar{c} \in F_0^\star$. Also

$$0 = Q(v) = Q(au + c^{-1}v) = (a + \bar{a})/c,$$

so $\bar{a} = -a$, i.e. $a \in \operatorname{Ker}\operatorname{tr}_{F/F_0}$. Thus σ is represented by $\begin{bmatrix} c & a \\ 0 & c^{-1} \end{bmatrix}$, and conversely each σ so represented is in $\operatorname{Stab}_{\operatorname{SU}(V)}([u])$. Since $\operatorname{SU}(V)$ is transitive on $\mathbb{P}\mathcal{C}$ and

$$|\mathbb{P}\mathcal{C}| = \frac{q^3 + (q^2 - q) - 1}{q^2 - 1} = \frac{(q^2-1)(q+1)}{q^2 - 1} = q + 1$$

we conclude that

$$|\operatorname{SU}(V)| = (q+1)|\operatorname{Stab}([u])| = (q+1)q(q-1).$$

Since $[\operatorname{U}(V) \colon \operatorname{SU}(V)] = q+1$ we conclude that $|\operatorname{U}(V)| = q(q+1)(q^2-1)$, as needed.

Now use induction on n and assume that $n \geq 3$. By Witt's Extension Theorem (10.12) $\mathrm{U}(V)$ acts transitively on the set of hyperbolic pairs in V. If (u,v) is a hyperbolic pair, and if $H = \langle u, v\rangle$, then clearly

$$\mathrm{Stab}_{\mathrm{U}(V)}((u,v)) = \mathrm{U}_{H^\perp}(V) \cong \mathrm{U}(H^\perp).$$

Thus

$$\begin{aligned} |\,\mathrm{U}(V)| &= h_n \cdot |\,\mathrm{U}(H^\perp)| \\ &= q^{2n-3}[q^{2n-1} + (-1)^n(q^n - q^{n-1}) - 1] \\ &\quad \cdot q^{\frac{(n-2)(n-3)}{2}}(q+1)(q^2-1)\cdots(q^{n-2}-(-1)^n) \\ &= q^{\frac{n^2-n}{2}} \cdot \prod_{j=1}^{n}(q^j - (-1)^j) \end{aligned}$$

by induction. □

Corollary 11.29. $|\,\mathrm{SU}(n,q^2)| = q^{\frac{n^2-n}{2}}(q^2-1)(q^3+1)\cdots(q^n-(-1)^n)$ *and* $|\,\mathrm{PSU}(n,q^2)| = |\,\mathrm{SU}(n,q^2)|/(n,q+1)$.

The final formula follows from the fact that

$$Z(\mathrm{SU}(n,q^2)) = \{a1 \colon a^n = a\bar{a} = 1\}.$$

Note in particular that if $|F| = 4$ then $|\mathbb{P}\mathcal{C}| = 3$ and $|\,\mathrm{PSU}(2,2^2)| = 6$, so $\mathrm{PSU}(2,2^2) \cong \mathrm{Sym}(3)$; if $|F| = 9$ then $|\mathbb{P}\mathcal{C}| = 4$ and

$$|\,\mathrm{PSU}(2,3^2)| = 3\cdot 2\cdot 4/(2,4) = 12,$$

so $\mathrm{PSU}(2,3^2) \cong \mathrm{Alt}(4)$.

Suppose that $n = 3$ and $q = 2$. Then $i_3 = 2^5 - (2^3 - 2^2) = 28$, so $|\mathbb{P}\mathcal{C}| = (28-1)/3 = 9$, and

$$|\,\mathrm{PSU}(3,2^2)| = 2^3(2^2-1)(2^3+1)/(3,3) = 72.$$

It is a consequence of Burnside's p^aq^b-Theorem (see e.g. [**28**] or [**30**]) that $\mathrm{PSU}(3,2^2)$ is solvable (in fact, the Sylow 3-subgroup is unique (hence normal); see the remarks below following the proof of Corollary 11.31).

Proposition 11.30. *If* $n = 3$ *and* $|F| = q^2$ *then* $\mathrm{PSU}(V)$ *is doubly transitive on* $\mathbb{P}\mathcal{C}$.

Proof. We know that $V = H \oplus \langle w\rangle$, with H a hyperbolic plane and $Q(w) = 1$. Choose a hyperbolic pair (u,v) in H and use $\{u,w,v\}$ as a basis. It will suffice to show that the stabilizer $\mathrm{Stab}_{\mathrm{SU}(V)}([u])$ is transitive on $\mathbb{P}\mathcal{C} \setminus \{[u]\}$. Note that

$$\widehat{B} = \begin{bmatrix} 0 & 0 & 1 \\ 0 & 1 & 0 \\ 1 & 0 & 0 \end{bmatrix}.$$

Suppose that $x \in V$ is isotropic, and write $x = au + bw + cv$. If $c = 0$ then also $b = 0$, in which case $[x] = [u]$. Thus if $[x] \in \mathbb{P}\mathcal{C} \setminus \{[u]\}$ we may assume that $x = au + bw + v$, and then $0 = Q(x) = a + \overline{a} + b\overline{b}$. Define σ via $\sigma u = u$, $\sigma w = -\overline{b}u + w$, and $\sigma v = au + bw + v$, i.e. σ is represented by

$$\begin{bmatrix} 1 & -\overline{b} & a \\ 0 & 1 & b \\ 0 & 0 & 1 \end{bmatrix}.$$

Check that $\sigma \in \mathrm{SU}(V)$; clearly $\sigma[u] = [u]$, and $\sigma[v] = [x]$, and the proof is complete. □

Corollary 11.31. $\mathrm{PSU}(3,4)$ *is sharply doubly transitive (see e.g.* [**30**] *for a definition) on* $\mathbb{P}\mathcal{C}$.

Proof. It is doubly transitive on $\mathbb{P}\mathcal{C}$, $|\mathbb{P}\mathcal{C}| = 9$, and $|\mathrm{PSU}(3,4)| = 9 \cdot 8$. □

In the case of $n = 3$ and $q^2 = 4$ write

$$M_{a,b} = \begin{bmatrix} 1 & -\overline{b} & a \\ 0 & 1 & b \\ 0 & 0 & 1 \end{bmatrix}.$$

Then the subgroup

$$M = \{M_{a,b} \colon a + \overline{a} + b\overline{b} = 0\}/Z(\mathrm{SU}(V))$$

stabilizes $[u]$ and has order 8 (it is a quaternion group), and therefore is of index 9 in $\mathrm{PSU}(3,4)$, so $M = \mathrm{Stab}_{\mathrm{PSU}}([u])$. The 9 conjugates of M are the stabilizers of the 9 points in $\mathbb{P}\mathcal{C}$. If 2 of those stabilizers had nontrivial intersection then a 2-element subset of $\mathbb{P}\mathcal{C}$ would have a nontrivial stabilizer, contradicting the sharp 2-transitivity of $\mathrm{PSU}(3,4)$. Thus

$$|\cup \{M^\sigma \colon \sigma \in \mathrm{PSU}(V)\}| = 1 + 9 \cdot 7 = 64,$$

so $\mathrm{PSU}(3,4) \setminus \cup\{M^\sigma \setminus \{1\} \colon \sigma \in \mathrm{PSU}(3,4)\}$ has cardinality $72 - 63 = 9$, and is therefore the unique Sylow 3-subgroup of $\mathrm{PSU}(3,4)$.

Chapter 12

Orthogonal Geometry (char $F = 2$)

If char $F \neq 2$ and V is an F-vector space recall that the quadratic form Q corresponding to a bilinear form B is defined by $Q(v) = B(v, v)$. If we define $\widetilde{Q}$ to be the scaled form $(1/2)Q$ then $\widetilde{Q}$ satisfies

$$\widetilde{Q}(ax + by) = a^2\widetilde{Q}(x) + abB(x, y) + b^2\widetilde{Q}(y)$$

for all $a, b \in F$, all $x, y \in V$, and consequently the bilinear form B is also easily recovered from the quadratic form $\widetilde{Q}$.

Suppose now, and for the rest of this chapter, that V is a vector space over a field F of characteristic 2, so the preceding paragraph does *not* apply. We *define* a *quadratic form* Q on V to be a function $Q\colon V \to F$ satisfying

$$Q(ax + by) = a^2Q(x) + abB(x, y) + b^2Q(y)$$

for all $a, b \in F$, all $x, y \in V$, where B is some bilinear form on V. Thus

$$B(x, y) = Q(x + y) + Q(x) + Q(y),$$

and the bilinear form is uniquely determined by Q. In particular

$$B(x, x) = Q(x + x) + Q(x) + Q(x) = 0,$$

all $x \in V$, so B must be an alternating form, and, since char $F = 2$, B is also symmetric. Furthermore Q is *not* uniquely determined by B.

If $\dim V = n$ then, since B is alternate, $\operatorname{rank} B = 2p$ and $\operatorname{rad}_B V$ is of dimension $n - 2p$ for some p. By Theorem 2.10 there is a basis

$$\{u_1, v_1, \ldots, u_p, v_p; w_1, \ldots, w_{n-2p}\}$$

for V, with each (u_i, v_i) a B-hyperbolic pair, $\operatorname{rad} V = \langle w_1, \dots, w_{n-2p} \rangle$, and

$$V = H_1 \oplus \cdots \oplus H_p \oplus \operatorname{rad}(V),$$

with $H_i = \langle u_i, v_i \rangle$. Thus $\widehat{B} = \operatorname{diag}(M, \dots, M, 0)$, with $M = \begin{bmatrix} 0 & 1 \\ 1 & 0 \end{bmatrix}$.

If $\operatorname{rad} V \neq 0$ say that V is *defective*; if $\operatorname{rad} V = 0$ say that V is *non-defective.* Note that "nondefective" describes the situation we previously called "nondegenerate". It is customary to use the term "nondegenerate" in a different sense in the current setting, although we shall not do so (see below).

All nonzero vectors $v \in V$ are B-isotropic, in the sense that $B(v, v) = 0$, but it does not generally follow that $Q(v) = 0$. If $0 \neq v \in V$ and $Q(v) = 0$ we say that v is a *singular* vector. A subspace W is called *singular* if there is a singular vector in W, otherwise W is *nonsingular*; W is *totally singular* if $Q(v) = 0$ for all $v \in W$. Note that if W is totally singular then

$$B(x, y) = Q(x + y) + Q(x) + Q(y) = 0$$

for all x, $y \in W$, so W is also totally isotropic relative to B.

Since char $F = 2$ the subset $F^{\square} = \{a^2 \colon a \in F\}$ is a subfield of F, and in fact the map $\alpha \colon F \to F^{\square}$ defined by $\alpha(a) = \overline{a} = a^2$ is an isomorphism from F to $F^{\square}$. Recall that if $F = F^{\square}$ then F is called a *perfect* field.

If x, $y \in \operatorname{rad} V$ and a, $b \in F$ note that $Q(ax + by) = a^2 Q(x) + b^2 Q(y) = \overline{a}Q(x) + \overline{b}Q(y)$, so Q restricts to an α-semilinear transformation from the F-vector space $\operatorname{rad} V$ to the $F^{\square}$-vector space F. We say that Q is *regular* on V if $\operatorname{Ker}(Q|_{\operatorname{rad} V}) = 0$ (this concept is sometimes called *nondegeneracy* in the literature).

Exercise

Suppose that F is a perfect field of characteristic 2 and Q is a regular quadratic form on a vector space V over F. Show that $\dim \operatorname{rad} V$ is 0 or 1.

A *quadratic space* at characteristic 2 is a vector space V over a field F of characteristic 2 having a regular quadratic form Q, with corresponding bilinear form B. We specifically allow the possibility that $\operatorname{rad}_B V \neq 0$, i.e. that V is *defective.*

If (V_1, Q_1) and (V_2, Q_2) are quadratic spaces over F then an *isometry* from V_1 to V_2 is an isomorphism $\sigma \colon V_1 \to V_2$ such that $Q_2(\sigma v) = Q_1(v)$ for all $v \in V_1$. We say that Q_1 and Q_2 are *equivalent* if there is an isometry from V_1 to V_2. If B_1 and B_2 are the bilinear forms determined by Q_1 and

Q_2, and if $\sigma\colon V_1 \to V_2$ is an isometry, then

$$\begin{aligned} B_2(\sigma x, \sigma y) &= Q_2(\sigma x + \sigma y) + Q_2(\sigma x) + Q_2(\sigma y) \\ &= Q_1(x + y) + Q_1(x) + Q_1(y) \\ &= B_1(x, y), \end{aligned}$$

so σ is also an isometry relative to B_1 and B_2. In particular $\sigma(\operatorname{rad} V_1) = \operatorname{rad} V_2$.

If V is a quadratic space with quadratic form Q then a *hyperbolic pair* relative to Q is an ordered pair (u, v) of singular vectors with $B(u, v) = 1$ (this is generally a stronger requirement, because of the singularity, than requiring (u, v) to be a B-hyperbolic pair). As usual if $\dim V = 2$ and V has a hyperbolic pair then V is called a *hyperbolic plane.*

Assume for the remainder of the chapter that V is a regular *quadratic space of dimension at least 2.*

Proposition 12.1. *If $u \in V \setminus \operatorname{rad} V$ is singular then there exists $v \in V$ so that (u, v) is a hyperbolic pair.*

Proof. Since $u \notin \operatorname{rad} V$ there exists $w \in V$ such that $B(u, w) \neq 0$; we may assume that $B(u, w) = 1$. Set $v = Q(w)u + w$. Then

$$Q(v) = Q(w)B(u, w) + Q(w) = 0,$$

and

$$B(u, v) = B(u, Q(w)u + w) = B(u, w) = 1.$$

□

Proposition 12.2. *If V is singular and nondefective then V has a basis of singular vectors.*

Proof. Choose a singular vector u. By Proposition 12.1 there exists a singular vector v such that (u, v) is a hyperbolic pair. Set $H = \langle u, v\rangle$, then $V = H \perp H^\perp$ since H is nondefective. Choose a basis $\{w_3, \dots, w_n\}$ for $H^\perp$, and set $W_i = \langle u, v + w_i\rangle$, $3 \leq i \leq n$. Then $B|_{W_i}$ is represented by the matrix $\begin{bmatrix} 0 & 1 \\ 1 & 0 \end{bmatrix}$, so $\operatorname{rad} W_i = 0$, and therefore W_i is regular. Thus by Proposition 12.1 there exist vectors $x_i \in W_i$ such that $Q(x_i) = 0$ and $B(u, x_i) = 1$. Since $w_i \in \langle u, v, x_i\rangle$ we have $V = \langle u, v, x_3, \dots, x_n\rangle$, and $\{u, v, x_3, \dots, x_n\}$ is a basis of singular vectors. □

Proposition 12.3. *Suppose that F is perfect,* $\dim V = n \geq 2$, *and V is defective. Then V is singular.*

Proof. By the exercise on page 114 $V \neq \operatorname{rad} V$. Choose $y \in \operatorname{rad} V$, $y \neq 0$, and $x \in V \setminus \operatorname{rad} V$. Then $Q(y) \neq 0$, since V is regular, and $Q(x)/Q(y) \in F = F^{\square}$, so $Q(x) = a^2 Q(y)$ for some $a \in F$. Set $u = x + ay \notin \operatorname{rad} V$ (since $x \notin \operatorname{rad} V$) and note that $Q(u) = Q(x) + a^2 Q(y) = 2Q(x) = 0$. □

Corollary 12.4. *There is a hyperbolic plane in V.*

Proposition 12.5. *Suppose that F is perfect, $\dim V = n \geq 3$, and V is nondefective. Then V is singular.*

Proof. Choose $x \neq 0$ in V and choose $y \in x^{\perp} \setminus \langle x \rangle$ ($\dim x^{\perp} \geq 2$). If $Q(x) = 0$ then x is singular. If $Q(x) \neq 0$ then $Q(y)/Q(x) \in F = F^{\square}$, so we may write $Q(y) = a^2 Q(x)$, $a \in F$, and set $u = ax + y \neq 0$. Then $Q(u) = a^2 Q(x) + Q(y) = 0$, so u is singular. □

Corollary 12.6. *There is a hyperbolic plane in V.*

Proposition 12.7. *Suppose that F is a perfect field. If V is a nondefective quadratic space over F that is not singular, with $\dim V = 2$, then there is a basis $\{v_1, v_2\}$ for V and an irreducible polynomial $x^2 + x + b \in F[x]$ such that*

$$Q(a_1 v_1 + a_2 v_2) = a_1^2 + a_1 a_2 + b a_2^2,$$

all $a_i \in F$.

Conversely, if $x^2 + x + b \in F[x]$ is irreducible and $\{v_1, v_2\}$ is a basis for V define $Q\colon V \to F$ via $Q(a_1 v_1 + a_2 v_2) = a_1^2 + a_1 a_2 + b a_2^2$, all $a_i \in F$. Then Q is a quadratic form for which V is nondefective and is not singular.

Proof. $\Rightarrow$: Since V is nondefective there exist $w_1, w_2 \in V$ with $B(w_1, w_2) = 1$. Thus $Q(a_1 w_1 + a_2 w_2) = a_1^2 Q(w_1) + a_1 a_2 + Q(w_2) a_2^2$, and $Q(w_1) \neq 0$ since V is not singular. Write $Q(w_1) = c^2 \in F^{\square} = F$, and set $v_1 = c^{-1} w_1$, $v_2 = c w_2$, and $b = c^2 Q(w_2)$. Then $\{v_1, v_2\}$ is a basis and

$$\begin{aligned} Q(a_1 v_1 + a_2 v_2) &= Q(a_1 c^{-1} w_1 + a_2 c w_2) \\ &= a_1^2 c^{-2} Q(w_1) + a_1 a_2 + a_2^2 c^2 Q(w_2) \\ &= a_1^2 + a_1 a_2 + b a_2^2. \end{aligned}$$

If $a \in F$ then $Q(a v_1 + v_2) = a^2 + a + b \neq 0$ since V is not singular, so $x^2 + x + b$ is irreducible. □

Exercise

Prove the converse in Proposition 12.7. It is not necessary to assume that F is perfect.

Proposition 12.8. *Suppose that V is a nondefective quadratic space but if V is a hyperbolic plane then $|F| > 2$. Then V has a basis of nonsingular vectors.*

Proof. There is nothing to prove if V is nonsingular, so assume that V is singular. Choose a hyperbolic pair (u, v) in V and set $H = \langle u, v\rangle$, so $V = H \oplus H^\perp$. If $H^\perp = 0$ then $|F| > 2$ and there exists $a \in F \setminus \{0, 1\}$. Then

$$\{w_1 = u + v,\ w_2 = u + av\}$$

is a basis of nonsingular vectors. Assume then that $H^\perp \neq 0$ and consider two cases.

1. Suppose that $H^\perp$ is not a hyperbolic plane over $\mathbb{F}_2$. By induction there is a nonsingular basis $\{w_3, \dots, w_n\}$ for $H^\perp$. Set $w_1 = u + w_3$, $w_2 = v + w_3$ and check that $\{w_1, \dots, w_n\}$ is a nonsingular basis for V.

2. Suppose that $F = \mathbb{F}_2$ and $H^\perp = \langle u_1, v_1\rangle$, (u_1, v_1) a hyperbolic pair. Set $w_1 = u + v + u_1$, $w_2 = u + v + v_1$, $w_3 = u + v$, $w_4 = v + u_1 + v_1$ and verify that $\{w_1, \dots, w_4\}$ is a nonsingular basis. □

Theorem 12.9. *Suppose that F is perfect and V is a quadratic space over F. Then V has a basis $\{v_i\}$ relative to which Q takes one of the following three forms.*

1. *If $n = 2m + 1$ is odd then*

$$Q(\sum_i a_i v_i) = a_1 a_{m+1} + a_2 a_{m+2} + \cdots + a_m a_{2m} + a_{2m+1}^2.$$

2. *If $n = 2m$ is even then either*
   (a) $Q(\sum_i a_i v_i) = \sum_{i=1}^m a_i a_{i+m}$, *or*
   (b) $Q(\sum_i a_i v_i) = \sum_{i=1}^{m-1} a_i a_{m-1+i} + a_{2m-1}^2 + a_{2m-1}a_{2m} + b a_{2m}^2$,
   with $x^2 + x + b$ irreducible in $F[x]$.

Proof. Since F is perfect and V is regular we have $\dim \operatorname{rad} V \leq 1$ by the exercise on page 114. We may successively apply Corollaries 12.4 and 12.6 and conclude that

$$V = H_1 \oplus H_2 \oplus \cdots \oplus H_m \oplus W,$$

where each H_i is a hyperbolic plane and $\dim W = 0$, 1, or 2. Let (v_i, v_{m+i}) be a hyperbolic pair in H_i. If $\dim W = 0$ we have Case 2(a). If $\dim W = 1$ then $W = \operatorname{rad} V = \langle v_{2m+1}\rangle$, and we may assume that $Q(v_{2m+1}) = 1$ since $F^\square = F$. We then have Case 1. If $\dim W = 2$, with W not singular, then Proposition 12.7 applies to W, and W has a basis $\{v_{2m-1}, v_{2m}\}$ relative to which we have Case 2(b). □

Theorem 12.10 (Witt's Extension Theorem). *Suppose that V has a nondefective quadratic form Q, that V_1 and V_2 are subspaces of V, and that $\sigma\colon V_1 \to V_2$ is an isometry. Then there is an isometry $\tau\colon V \to V$ such that $\tau|_{V_1} = \sigma$.*

Proof. The proof of Theorem 10.12 carries over, with small changes. The following paragraphs should be substituted for the corresponding sections in the proof of 10.12.

(For (*):) Then

$$\begin{aligned} Q(\tau v) &= Q(\sigma v_1 + x) \\ &= Q(\sigma v_1) + B(\sigma v_1, x) + Q(x) \\ &= Q(v_1) + B(v_1, x) + Q(x) \\ &= Q(v_1 + x) = Q(v), \end{aligned}$$

and τ is an isometry.

(For (**):) Observe that

$$\begin{aligned} Q(\sigma v_0 - v_0) &= Q(\sigma v_0) + B(\sigma v_0, v_0) + Q(v_0) \\ &= B(\sigma v_0, v_0) + B(v_0, v_0) = 0 \end{aligned}$$

since $v_0 \in V_1 \subseteq W^\perp = \langle \sigma v_0 - v_0 \rangle$. Thus W is a (totally) singular line, and $W \subseteq W^\perp$.

(For (***):) Thus $Q(u + aw_0) = Q(u) + aB(u, w_0)$. Take

$$a = (Q(u) + Q(v))/B(u, w_0).$$

Then $Q(u + aw_0) = Q(u) + Q(u) + Q(v) = Q(v)$. □

Corollary 12.11. *Any two maximal totally singular subspaces of V are isometric. The common dimension of all of them is called the* Witt index $m = m(V)$.

Remark

It follows easily from Proposition 12.1, by splitting off hyperbolic planes, that $m \leq \lfloor n/2 \rfloor$.

Corollary 12.12 (Witt's Cancellation Theorem). *Suppose that V is a non-defective quadratic space and that V_1, V_2 are isometric subspaces of V. Then $V_1^\perp$ and $V_2^\perp$ are also isometric.*

Chapter 13

Clifford Algebras (char $F = 2$)

Assume for this chapter that char $F = 2$ *and that* V *is a vector space over* F *with a quadratic form* Q. *Thus* V *has a corresponding alternate and symmetric bilinear form* B *given by*

$$B(x, y) = Q(x + y) + Q(x) + Q(y).$$

The Clifford algebra $C = C(V)$ is defined and constructed exactly as in Chapter 8: $C(V) = T(V)/I$, where I is the ideal in the tensor algebra $T(V)$ generated by all $v \otimes v - Q(v) \cdot 1$, $v \in V$. As before we may assume that F is a subalgebra of $C(V)$ and V is a subspace of $C(V)$.

By virtually the same proofs we have analogues of the first four results of Chapter 8, as rephrased on page 69.

Proposition 13.1. *If* $u, v \in V$ *then* $B(u, v) = uv + vu$.

Proposition 13.2. *If* $u, v \in V$ *then* $u \perp v$ *if and only if* $uv = vu$.

Proposition 13.3. *If* $0 \neq v \in V$ *then* v *is a unit in* C *if and only if* v *is nonsingular in* V.

Proposition 13.4. *If* $u \in V$ *is nonzero and nonsingular then* $\sigma_u w = uwu^{-1}$ *for all* $w \in V$.

Proposition 13.5. *If* $\{v_1, \ldots, v_n\}$ *is a basis for* V *then*

$$\{v_1^{e_1} v_2^{e_2} \cdots v_n^{e_n} : \text{ each } e_i = 0 \text{ or } 1\}$$

is a spanning set for $C = C(V)$ *as a vector space. Thus* $\dim_F C \leq 2^n$.

Proof. V generates $C(V)$ as an algebra, so every $x \in C(V)$ is a linear combination of products of v_i's. Given any product of v_i's apply $v_i^2 = Q(v_i)$ and $v_i v_j = v_j v_i + B(v_i, v_j)$ (Proposition 13.1) to rewrite the product as a linear combination of products each of the form $v_1^{e_1} v_2^{e_2} \cdots v_n^{e_n}$, with each e_i being 0 or 1. □

As in Chapter 8 $C(V)$ is $\mathbb{Z}_2$-graded, $C(V) = C_0(V) \oplus C_1(V)$, with $C_0(V)$ the *even subalgebra.*

Since char $F = 2$ the "twisted tensor product" of algebras, as discussed in Chapter 8, is simply the usual tensor product, and thus if $V = V_1 \perp V_2$ then $C(V) = C(V_1) \otimes C(V_2)$.

Exercises

1. If $V = \langle v \rangle$, a space of dimension 1, with $Q(v) = a \in F$, show that either $C(V) \cong F(\sqrt{a})$, a field, or $C(V) \cong F[x]/\langle (x+b)^2 \rangle$, $b^2 = a$, in which case $C(V)$ is not semisimple. In either case $\dim C(V) = 2$.
2. Suppose that V is a hyperbolic plane, with hyperbolic pair (u, v). Define

$$\varphi(u) = \begin{bmatrix} 0 & 1 \\ 0 & 0 \end{bmatrix}, \quad \varphi(v) = \begin{bmatrix} 0 & 0 \\ 1 & 0 \end{bmatrix} \in M_2(F),$$

and show that φ extends to an algebra isomorphism between $C(V)$ and $M_2(F)$, so $\dim C(V) = 4$.

Let us generalize Exercise 2 above. Given a, $b \in F$ define a *Huppert algebra* [**42**] (analogous to a *(generalized) quaternion algebra*, see e.g [**45**] or [**58**]) over F to be a 4-dimensional vector space A having a basis $\{1, i, j, k\}$, with multiplication defined as follows. The basis element 1 is the multiplicative identity of A, and we identity F with the subalgebra $F \cdot 1$ of A. Beyond that the multiplication is determined by the following multiplication table.

	i	j	k
i	a	k	aj
j	$1+k$	b	$bi+j$
k	$i+aj$	bi	$ab+k$

An easy, but somewhat tedious, calculation shows that A is an associative algebra (e.g. $(ij)k = k^2 = ab + k$, whereas $i(jk) = i(bi + j) = ab + k$, etc.). Notation: we write $A = H(a, b)$.

Proposition 13.6. *If V is a nondefective quadratic space of dimension 2 then $C(V)$ is a Huppert algebra, and hence* $\dim C(V) = 4$.

Proof. Choose a basis $\{u, v\}$ for V with $B(u, v) = 1$, and say $Q(u) = a$, $Q(v) = b$. Set $A = H(a, b)$. The linear transformation $\varphi\colon V \to A$ determined by $\varphi(u) = i$, $\varphi(v) = j$ is compatible with Q, so φ extends to an algebra homomorphism $\tilde{\varphi}\colon C(V) \to A$. Since $\{i, j\}$ generates A as an algebra $\tilde{\varphi}$ is surjective. By Proposition 13.5 we have

$$4 \geq \dim C(V) \geq \dim A = 4$$

and $C(V) \cong A$. □

Remark

If $a = b = 0$ in the proof above then it is easy to check that $H(0, 0) \cong M_2(F)$, in agreement with Exercise 2 above.

Recall that an algebra A over a field F is called *central* if its center is $Z(A) = \{a \cdot 1\colon a \in F\} \cong F$.

Proposition 13.7. *If a, $b \in F$ then the Huppert algebra $A = H(a, b)$ is a central simple algebra.*

Proof. If x, $y \in A$ write $[x, y] = xy + yx$, and write $x = c_0 + c_1 i + c_2 j + c_3 k$. Check that $[x, i] = c_2 + c_3 i$, $[x, j] = c_1 + c_3 j$, and $[[x, i], j] = c_3$. If $x \in Z(A)$ then $[x, i] = [x, j] = 0$ and hence $c_1 = c_2 = c_3 = 0$, and $x = c_0 \in F$, so A is central.

Suppose that $I \neq A$ is an ideal in A and take $x = c_0 + c_1 i + c_2 j + c_3 k \in I$. Since $c_3 = [[x, i], j] \in I$ we have $c_3 = 0$, hence $c_2 = [x, i]$, $c_1 = [x, j] \in I$, and therefore $c_1 = c_2 = 0$. Finally then $x = c_0 \in I$, so $x = 0$ and A is simple. □

Proposition 13.8. *If V is a quadratic space of dimension n then $C(V)$ has dimension 2^n.*

Proof. Induction on n. If $n = 1$ the result is Exercise 1 on page 120, so assume that $n \geq 2$ and assume the result for lower dimensions.

Suppose first that V is defective and choose $u \in \operatorname{rad} V \setminus \{0\}$. Choose a complementary subspace W orthogonal to $\langle v \rangle$, so $V = \langle v \rangle \textcircled{\perp} W$. Then $C(V) \cong C(\langle v \rangle) \otimes C(W)$, and by Exercise 1 on page 120 and induction we have $\dim C(V) = 2 \cdot 2^{n-1} = 2^n$.

Suppose next that V is nondefective, and choose u, $v \in V$ with $B(u, v) = 1$. Set $U = \langle u, v \rangle$, so $V = U \textcircled{\perp} U^{\perp}$, $C(V) \cong C(U) \otimes C(H^{\perp})$. By Proposition 13.6 and induction we have $\dim C(V) = 4 \cdot 2^{n-2} = 2^n$. □

Corollary 13.9. *If $\{v_1, v_2, \ldots, v_n\}$ is a basis for V then*

$$\{v_1^{e_1} v_2^{e_2} \cdots v_n^{e_n} \colon \text{each } e_i = 0 \text{ or } 1\}$$

is a basis for $C(V)$.

Corollary 13.10. $\dim C_0(V) = \dim C_1(V) = 2^{n-1}$.

Proposition 13.11. *Suppose that V is a nondefective quadratic space of even dimension $n \geq 2$. Then the Clifford algebra $C(V)$ is central simple.*

Proof. Induction. For $n = 2$ see Propositions 13.6 and 13.7, so we may assume that $n \geq 4$ and assume the result for lower dimensions. Choose $u, v \in V$ with $B(u,v) = 1$ and set $W = \langle u, v\rangle$. Then $C(V) \cong C(W) \otimes C(W^{\perp})$. Since $C(W)$ is central simple by Propositions 13.6 and 13.7, and $C(W^{\perp})$ is central simple by induction, $C(V)$ is also central simple (see e.g. [**58**], page 225). □

We begin now a somewhat technical discussion aimed at expanding on the description of possible quadratic forms over perfect fields begun in Theorem 12.9.

Remark

If V is a quadratic space there are nonzero nonsingular vectors in V. This is immediate if $\operatorname{rad} V \neq 0$ since Q is regular on V. If $\operatorname{rad} V = 0$ choose $x, y \in V$ with $B(x, y) = 1$. Then $Q(x+y) + Q(x) + Q(y) = 1$, so at least one of x, y, and $x + y$ is nonzero and nonsingular.

If V is a quadratic space choose $u_1 \in V$ with $Q(u_1) \neq 0$. Given any $u, v \in V$ we have

$$(u_1u)(u_1v) = u_1[u_1u + B(u, u_1)]v = Q(u_1)uv + B(u, u_1)u_1v$$

in $C(V)$ by Proposition 13.1. Thus

$$uv = \frac{1}{Q(u_1)}[u_1u + B(u, u_1)]u_1v,$$

and it follows that the even subalgebra $C_0(V)$ is generated by the set of all elements u_1u, as u ranges over V.

Keep u_1 fixed (with $Q(u_1) \neq 0$), and let C' be the subalgebra of $C_0(V)$ generated by all u_1v, $v \in u_1^{\perp}$. Since B is alternate $u_1 \in u_1^{\perp}$, and we may choose a basis $\{u_1, w_1, \ldots, w_{n-2}\}$ for $u_1^{\perp}$. Set $W = \langle w_1, \ldots, w_{n-2}\rangle$, an $n-2$ dimensional subspace of $u_1^{\perp}$, and note that C' is generated as an algebra by all u_1w_i, $1 \leq i \leq n-2$ (also $u_1^{\perp} = \langle u_1\rangle \bigoplus W$).

Define a quadratic form Q_1 on W by setting $Q_1(w) = Q(u_1)Q(w)$, all $w \in W$. Determine a map $\varphi\colon W \to C'$ by setting $\varphi(w_i) = u_1w_i$, $1 \leq i \leq n-2$. Then $\varphi(w_i)^2 = (u_1w_i)^2 = u_1^2w_i^2$ (since $w_i \perp u_1$), and thus $\varphi(w_i)^2 = Q(u_1)Q(w_i) = Q_1(w_i)$, all i. In other words C' is compatible, via φ, with Q_1 on W, and we may extend φ to an algebra homomorphism from $C(W, Q_1)$ onto C'.

Note that C' contains $(u_1w_i)(u_1w_j) = Q(u_1)w_iw_j$ for all i, j. It follows easily that C' contains all $w_1^{e_1}w_2^{e_2}\cdots w_{n-2}^{e_{n-2}}$, where each e_i is 0 or 1 and $\sum_i e_i$ is even, and also all $u_1w_1^{e_1}\cdots w_{n-2}^{e_{n-2}}$, with $\sum_i e_i$ odd. Those are linearly independent by Corollary 13.9, and so $\dim C' \geq 2^{n-2}$. Thus

$$2^{n-2} = \dim C(W, Q_1) \geq \dim C' \geq 2^{n-2},$$

and we have $C' \cong C(W, Q_1)$.

Suppose now that V is a nondefective quadratic space, so $\dim V = n = 2k$ is even. Choose a B-symplectic basis $\{u_1, v_1, \ldots, u_k, v_k\}$, with $Q(u_1) \neq 0$ as above. By the construction above $C_0(V)$ has a subalgebra C' of dimension 2^{n-2} which is isomorphic with the Clifford algebra $C(W, Q_1)$ of a subspace W, and hence C' is central simple by Proposition 13.11.

Write $\mathcal{B}$ for the basis $\{u_i, v_i\}$ above. Set $z = z_{\mathcal{B}} = \sum_{i=1}^k u_iv_i \in C_0(V)$, and set $\mathfrak{a} = \mathfrak{a}_{\mathcal{B}} = \sum_{i=1}^k Q(u_i)Q(v_i) \in F$. Note that all u_i commute, all v_j commute, and all u_i and v_j commute, except that $u_iv_i = v_iu_i + 1$ for each i. As a consequence it is easily checked that $zu_i = u_iz + u_i$, $zv_j = v_jz + v_j$, all i and j. It follows that z commutes with a set of generators for $C_0(V)$ (e.g. $zu_iv_j = (zu_i + u_i)v_j = u_iv_jz + u_iv_j + u_iv_j = u_iv_jz$), and hence $z \in Z(C_0)$. Clearly $z \notin F$, so $\dim Z(C_0) \geq 2$.

If $i \neq j$ then $u_iv_iu_jv_j = u_jv_ju_iv_i$, and the corresponding terms cancel in the expansion of $z^2 = \sum_{i,j} u_iv_iu_jv_j$. As a result we have

$$\begin{aligned} z^2 &= \sum_i u_iv_iu_iv_i = \sum_i u_i(u_iv_i + 1)v_i \\ &= \sum_i u_i^2v_i^2 + \sum_i u_iv_i = \mathfrak{a} + z. \end{aligned}$$

Since $\mathfrak{a} \in F$ the subalgebra of $C_0(V)$ generated by z is just the 2-dimensional subspace $C'' = F + Fz$, which is central in $C_0(V)$.

Keep the notation established above in the next theorem. Write $F^{[2]}$ for the additive subgroup $\{a^2 + a\colon a \in F\}$ of F.

Theorem 13.12. *Suppose that V is a nondefective quadratic space. Then $Z(C_0(V)) = F + Fz$, of dimension 2. If $\mathfrak{a} \notin F^{[2]}$ then $Z(C_0) = F(z)$ is a field extension of F and $C_0(V)$ is a simple algebra; if $\mathfrak{a} \in F^{[2]}$ then $Z(C_0(V)) \cong F \oplus F$ and $C_0(V) \cong C' \oplus C'$, a direct sum of two isomorphic central simple algebras.*

Proof. We have C'' contained in the centralizer of C', C' central simple, and $\dim C' \cdot \dim C'' = \dim C_0(V)$. It follows from Proposition (a) on page 225 of [**58**] that $C_0(V) \cong C' \otimes C''$ and hence $C'' = Z(C_0(V))$. Recall that $z^2 + z = \mathfrak{a}$. If $\mathfrak{a} \notin F^{[2]}$ then the polynomial $f(x) = x^2 + x + \mathfrak{a}$ is irreducible in $F[x]$. Since $z \in C''$ is a root of $f(x)$ we have $C'' = F(z)$, a quadratic field extension of F. In that case C'' is a simple F-algebra, and consequently $C_0(v) \cong C' \otimes C''$ is simple. If $\mathfrak{a} \in F^{[2]}$ say $\mathfrak{a} = b^2 + b$, $b \in F$. Set $u = b + z$ and $v = u + 1 = b + 1 + z \in C''$. Note that $u^2 = b^2 + z^2 = b^2 + \mathfrak{a} + z = b^2 + b^2 + b + z = u$, $v^2 = u^2 + 1 = u + 1 = v$, $uv = u(u+1) = u^2 + u = 2u = 0 \quad (= vu)$, and that $u + v = 1$. It follows easily that $C'' = Fu \oplus Fv \cong F \oplus F$, and so

$$C_0(V) \cong C' \otimes C'' \cong C' \otimes (F \oplus F) \cong C' \oplus C'.$$

□

Theorem 13.13. *Suppose that V is a nondefective quadratic space, and let $\{u_1, v_1, \ldots, u_m, v_m\}$ be a symplectic basis. Then any other symplectic basis can be obtained by choosing $\sigma \in \mathrm{Sp}_B(V)$ and taking $\{\sigma u_i, \sigma v_i\}$. As above define $z = \sum_i u_i v_i$, $\mathfrak{a} = \sum_i Q(u_i)Q(v_i)$, and $z' = \sum_i \sigma(u_i)\sigma(v_i)$, $\mathfrak{a}' = \sum_i Q(\sigma(u_i))Q(\sigma(v_i))$. Then there exists $\delta = \delta(\sigma) \in F$ such that $z' = z + \delta$ and $\mathfrak{a}' = \mathfrak{a} + \delta^2 + \delta$. If σ is an isometry relative to Q then $\delta = 0$ or 1.*

Proof. By Theorem 13.12 and the preceding discussion we have $z^2 + z = \mathfrak{a} \in F$, $z'^2 + z' = \mathfrak{a}' \in F$, and $Z(C_0(V)) = F + Fz = F + Fz'$. Thus $z' = bz + \delta$ for some $b, \delta \in F$, and so $z'^2 + z' = b^2 z^2 + \delta^2 + bz + \delta = b^2(z^2 + z) + \delta^2 + \delta + (b^2 + b)z \in F$. It follows that $b^2 + b$, the coefficient of z, is 0, and therefore $b = 0$ or 1. But $b \neq 0$ or else $z' = \delta \in F$, so $b = 1$ and therefore $z' = z + \delta$. Thus $\mathfrak{a}' = z'^2 + z' = (z+\delta)^2 + z + \delta = z^2 + z + \delta^2 + \delta = \mathfrak{a} + (\delta^2 + \delta)$. If σ is a Q-isometry then $\mathfrak{a}' = \mathfrak{a}$, so $\delta^2 + \delta = 0$, and $\delta = 0$ or 1. □

Recall that $F^{[2]}$ denotes the additive subgroup $\{a^2 + a \colon a \in F\}$ of F. By Theorem 13.13 the coset $\mathfrak{a} + F^{[2]} \in F/F^{[2]}$ is independent of the choice of symplectic basis in a nondefective quadratic space V. It is called the *Arf invariant* of V.

Note that the map $a \mapsto a^2 + a$ is a homomorphism from F onto $F^{[2]}$ with kernel the prime subfield $\mathbb{F}_2 = \{0, 1\} \leq F$. In particular, if F is finite then $[F \colon F^{[2]}] = 2$.

We may now complete the discussion of quadratic spaces over perfect fields of characteristic 2 discussed earlier in Theorem 12.9. It is clear from Theorem 12.9 that all quadratic forms of the same odd dimension are equivalent.

Proposition 13.14. *Suppose that F is perfect. In the context of Theorem 12.9 and its proof let V_1, V_2, and V_3 be quadratic spaces with respective quadratic forms Q_1 of type 2(a), and Q_2, Q_3 of types 2(b) relative to irreducible quadratics $x^2 + x + b_2$, $x^2 + x + b_3$, respectively, in $F[x]$. Then V_1 is not equivalent to V_2 (nor to V_3); V_2 and V_3 are equivalent if and only if $b_2 \equiv b_3$ (mod $F^{[2]}$).*

Proof. In Theorem 12.9 Q_1 is evaluated relative to a symplectic basis arranged as $\{u_1, \dots, u_m, v_1, \dots, v_m\}$ with $Q(u_i) = Q(v_j) = 0$, all i, j, so the Arf invariant is $\mathfrak{a}_1 \equiv 0 \pmod{F^{[2]}}$. Similarly $\mathfrak{a}_2 \equiv b_2 \pmod{F^{[2]}}$ and $\mathfrak{a}_3 \equiv b_3 \pmod{F^{[2]}}$. If V_1 and V_2 were equivalent then $\mathfrak{a}_1 \equiv \mathfrak{a}_2 \pmod{F^{[2]}}$, or $b_2 \equiv 0 \pmod{F^{[2]}}$, and therefore $b_2 = a^2 + a$ for some $a \in F$, contradicting the irreducibility of $x^2 + x + b_2$.

If V_2 and V_3 are equivalent then $\mathfrak{a}_2 = \mathfrak{a}_3$, i.e. $b_2 \equiv b_3 \pmod{F^{[2]}}$. Suppose conversely that $b_2 \equiv b_3 \pmod{F^{[2]}}$, say via $b_2 - b_3 = c^2 + c \in F^{[2]}$. We may assume that $\dim V_2 = \dim V_3 = 2$ and that $V_2 = \langle u_2, v_2 \rangle$, $V_3 = \langle u_3, v_3 \rangle$, relative to which Q_2 and Q_3 have the respective forms $a_1^2 + a_1 a_2 + b_2 a_2^2$, $a_1^2 + a_1 a_2 + b_3 a_2^2$, i.e. $Q_2(u_2) = 1$, $Q_2(v_2) = b_2$, $Q_3(u_3) = 1$, and $Q_3(v_3) = b_3$. Define $\sigma\colon V_2 \to V_3$ via $\sigma(u_2) = u_3$, $\sigma(v_2) = cu_3 + v_3$. Then

$$Q_3(\sigma u_2) = Q_3(u_3) = 1 = Q_2(u_2)$$

and

$$Q_3(\sigma v_2) = Q_3(cu_3 + v_3) = c^2 + c + b_3 = Q_2(2_2),$$

so σ is an isometry and therefore V_2 and V_3 are equivalent. □

Corollary 13.15. *If F is finite and $\dim V$ is even there are exactly 3 equivalence classes of quadratic forms on V.*

Proof. $[F\colon F^{[2]}] = 2$. □

Proposition 13.16. *If σ is an isometry of V then there exists a unique automorphism $\widehat{\sigma}$ of $C(V)$ such that $\widehat{\sigma}|_V = \sigma$.*

Proof. View σ as a linear transformation from V to $C(V)$. Then $\sigma(v)^2 = Q(\sigma v) = Q(v)$, so σ extends to a unique algebra homomorphism $\widehat{\sigma}$ from $C(V)$ to $C(V)$, and $\widehat{\sigma}$ is surjective since its image contains $\sigma V = V$, which generates $C(V)$. Thus $\widehat{\sigma}$ is an automorphism. □

Remark

Proposition 13.16 does not depend on F having characteristic 2.

Chapter 14

Orthogonal Groups (char $F = 2$)

In this chapter results about orthogonal groups from earlier chapters are either extended to or differentiated from corresponding results about groups of orthogonal transformations over fields of characteristic 2.

Throughout the chapter F is a field of characteristic 2 and V is a quadratic space over F, with (regular) quadratic form Q and associated bilinear form B.

The *orthogonal group* $\mathrm{O}(V)$ is the group of all Q-isometries of V. Note that

$$\mathrm{O}(V) \leq \mathrm{Sp}_B(V) = \mathrm{Sp}(V) \leq \mathrm{SL}(V),$$

so $\det \sigma = 1$, all $\sigma \in \mathrm{O}(V)$.

Theorem 14.1. *Let $V_0 = \operatorname{rad} V$ and let V_1 be a complement to V_0 in V, so that $V = V_0 \oplus V_1$.*

1. *If $\sigma \in \mathrm{O}(V)$ then $\sigma|_{V_0} = 1_{V_0}$ and $\sigma|_{V_1} = \sigma_0 + \sigma_1$, where $\sigma_0 \colon V_1 \to V_0$ and $\sigma_1 \in \mathrm{Sp}(V_1)$.*
2. *If $\tau \in \mathrm{Sp}(V_1)$ then there exists $\sigma \in \mathrm{O}(V)$ with $\sigma_1 = \tau$ if and only if*

$$Q(\tau v) + Q(v) \in Q(V_0)$$

for all $v \in V_1$. If such a σ exists it is unique.

3. *The map $\sigma \mapsto \sigma_1$ is a monomorphism from $\mathrm{O}(V)$ into $\mathrm{Sp}(V_1)$.*

Proof. Since $\sigma \in \mathrm{O}(V)$ is a symplectic transformation $\sigma V_0 = V_0$.

1. If $v \in V_0$ then $\sigma v + v \in V_0$, and $Q(\sigma v + v) = Q(\sigma v) + Q(v) = 2Q(v) = 0$, and therefore $\sigma v + v = 0$, or $\sigma v = v$, since V is regular. Thus $\sigma|_{V_0} = 1_{V_0}$. For each $v \in V_1$ write $\sigma v = \sigma_0 v + \sigma_1 v$, with $\sigma_i v \in V_i$, thereby defining functions $\sigma_i \colon V_1 \to V_i$, $i = 0, 1$. Clearly σ_0 and σ_1 are linear transformations. If $u, v \in V_1$ then

$$\begin{aligned} B(\sigma_1 u, \sigma_1 v) &= B(\sigma u + \sigma_0 u, \sigma v + \sigma_0 v) \\ &= B(\sigma u, \sigma v) = B(u, v) \end{aligned}$$

since $\sigma_0 u$, $\sigma_0 v \in \operatorname{rad} V$ and $\sigma \in \mathrm{O}(V)$, and so $\sigma_1 \in \mathrm{Sp}(V_1)$.

2. $\Rightarrow$: Suppose that $\tau = \sigma_1$ for some $\sigma \in \mathrm{O}(V)$. Since $\sigma_0 v \perp \sigma_1 v$, all $v \in V_1$, we have

$$Q(v) = Q(\sigma v) = Q(\sigma_0 v + \sigma_1 v) = Q(\sigma_0 v) + Q(\sigma_1 v),$$

and therefore

$$Q(\tau v) + Q(v) = Q(\sigma_1 v) + Q(v) = Q(\sigma_0 v) \in Q(V_0).$$

$\Leftarrow$: Assume that $Q(\tau v) + Q(v) \in Q(V_0)$ for all $v \in V_1$. Given $v \in V_1$ write $Q(\tau v) + Q(v) = Q(v_0)$, with $v_0 \in V_0$. Then v_0 is unique, for suppose that also $Q(\tau v) + Q(v) = Q(w_0)$, $w_0 \in V_0$. Then $0 = Q(w_0) - Q(v_0) = Q(w_0 - v_0)$, and therefore $w_0 = v_0$ since V is regular. Thus we may define $\sigma_0 \colon V_1 \to V_0$ via $\sigma_0 v = v_0$. It is left as an exercise to confirm that σ_0 is a linear transformation. Now define $\sigma \colon V \to V$ via $\sigma|_{V_0} = 1_{V_0}$ and $\sigma|_{V_1} = \sigma_0 + \tau$. If $v \in V$ write $v = v_0 + v_1$, $v_i \in V_i$. Then $\sigma v = v_0 + \sigma_0 v_1 + \tau v_1$, so

$$Q(\sigma v) = Q(v_0 + \sigma_0 v_1 + \tau v_1) = Q(v_0) + Q(\sigma_0 v_1) + Q(\tau v_1),$$

where $\sigma_0 v_1$ is the unique element of V_0 such that $Q(\tau v_1) + Q(v_1) = Q(\sigma_0 v_1)$. Thus

$$\begin{aligned} Q(\sigma v) &= Q(v_0) + Q(\tau v_1) + Q(v_1) + Q(\tau v_1) \\ &= Q(v_0) + Q(v_1) = Q(v_0 + v_1) = Q(v). \end{aligned}$$

Since 1_{V_0} and τ are isomorphisms σ is an isomorphism, and therefore $\sigma \in \mathrm{O}(V)$ with $\sigma_1 = \tau$. Uniqueness will be shown in the proof of part 3.

3. Take σ, $\sigma' \in \mathrm{O}(V)$, and write $\sigma = \sigma_0 + \sigma_1$, $\sigma' = \sigma_0' + \sigma_1'$ as in part 1. If $v \in V$ then

$$\begin{aligned} \sigma' \sigma v &= \sigma'(\sigma_0 v + \sigma_1 v) \\ &= \sigma_0 v + \sigma'(\sigma_1 v) \\ &= \sigma_0 v + \sigma_0' \sigma_1 v + \sigma_1' \sigma_1 v, \end{aligned}$$

and we see that $(\sigma'\sigma)_0 = \sigma_0 + \sigma_0'\sigma_1$, $(\sigma'\sigma)_1 = \sigma_1'\sigma_1$. In particular the map $\sigma \mapsto \sigma_1$ is a homomorphism from $\mathrm{O}(V)$ to $\mathrm{Sp}(V_1)$. Suppose $\sigma_1 = 1_{V_1}$. For any $v \in V$ write $v = v_0 + v_1$, $v_i \in V_i$. Then $\sigma v_1 = \sigma_0 v_1 + v_1$, and

$$Q(v_1) = Q(\sigma v_1) = Q(\sigma_0 v_1 + v_1) = Q(\sigma_0 v_1) + Q(v_1),$$

so $Q(\sigma_0 v_1) = 0$, therefore $\sigma_0 v_1 = 0$ by regularity, and $\sigma_0 \equiv 0$. Thus $\sigma = 1_V$ and $\sigma \mapsto \sigma_1$ is a monomorphism. The uniqueness in part 2 follows immediately. □

Theorem 14.2. *If F is perfect and V is a quadratic space of odd dimension $n = 2m+1$ then $\mathrm{O}(V) \cong \mathrm{Sp}(2m, F)$.*

Proof. Since n is odd V is defective and $\dim \operatorname{rad} V = 1$ by the Exercise on page 114. Say $\operatorname{rad} V = V_0 = \langle v_0 \rangle$. If $Q(v_0) = a$ ($\neq 0$) write $a = b^2$, so that $Q(b^{-1}v_0) = 1$. Thus $Q(V_0) = F^{\square} = F$. It follows from Theorem 14.1, Part 2, that the map $\sigma \mapsto \sigma_1$ is surjective for any complement V_1 to V_0 in V, and hence $\mathrm{O}(V) \cong \mathrm{Sp}(V_1)$. □

Assume until further notice that V is nondefective.

Recall from Chapter 13 that if $\mathcal{B} = \{u_i, v_i\}$ is a symplectic basis for V then $Z(C_0(V)) = F \oplus F z_{\mathcal{B}}$, with $z_{\mathcal{B}} = \sum_i u_i v_i$. If $\sigma \in \mathrm{Sp}(V)$ then $z_{\sigma\mathcal{B}} = z_{\mathcal{B}} + \delta_{\mathcal{B}}(\sigma)$ by Theorem 13.13.

Suppose that σ, $\tau \in \mathrm{Sp}(V)$. Then

$$\begin{aligned} z_{\sigma\tau\mathcal{B}} &= z_{\mathcal{B}} + \delta_{\mathcal{B}}(\sigma\tau) \\ &= z_{\sigma(\tau\mathcal{B})} = z_{\tau\mathcal{B}} + \delta_{\tau\mathcal{B}}(\sigma) \\ &= z_{\mathcal{B}} + \delta_{\mathcal{B}}(\tau) + \delta_{\tau\mathcal{B}}(\sigma), \end{aligned}$$

and consequently $\delta_{\mathcal{B}}(\sigma\tau) = \delta_{\mathcal{B}}(\tau) + \delta_{\tau\mathcal{B}}(\sigma)$.

Proposition 14.3. *If $\sigma \in \mathrm{O}(V)$ and $\tau \in \mathrm{Sp}(V)$ then $\delta_{\mathcal{B}}(\sigma\tau) = \delta_{\mathcal{B}}(\sigma) + \delta_{\mathcal{B}}(\tau)$.*

Proof. By Proposition 13.16 there is an automorphism $\widehat{\sigma}$ of $C(V)$ such that $\widehat{\sigma}|_V = \sigma$. Thus $\widehat{\sigma} z_{\mathcal{B}} = \sum_i (\sigma u_i)(\sigma v_i) = z_{\sigma\mathcal{B}}$, and of course

$$\widehat{\sigma}(a) = \widehat{\sigma}(a \cdot 1) = a \cdot \widehat{\sigma}(1) = a,$$

all $a \in F$. Thus $\widehat{\sigma} z_{\tau\mathcal{B}} = z_{\sigma\tau\mathcal{B}} = z_{\mathcal{B}} + \delta_{\mathcal{B}}(\sigma\tau)$, but also

$$\widehat{\sigma} z_{\tau\mathcal{B}} = \widehat{\sigma}(z_{\mathcal{B}} + \delta_{\mathcal{B}}(\tau)) = z_{\sigma\mathcal{B}} + \delta_{\mathcal{B}}(\tau) = z_{\mathcal{B}} + \delta_{\mathcal{B}}(\sigma) + \delta_{\mathcal{B}}(\tau),$$

and therefore $\delta_{\mathcal{B}}(\sigma\tau) = \delta_{\mathcal{B}}(\sigma) + \delta_{\mathcal{B}}(\tau)$. □

Corollary 14.4. *The restriction of $\delta_{\mathcal{B}}$ to $\mathrm{O}(V)$ is a homomorphism from $\mathrm{O}(V)$ to the additive group of the prime field $\mathbb{F}_2 = \{0, 1\} \subseteq F$.*

Proof. Theorem 13.13. □

The function $\delta = \delta_{\mathcal{B}} \colon \mathrm{O}(V) \to \mathbb{F}_2$ is called *Dickson's pseudodeterminant.*

Proposition 14.5. *The pseudodeterminant $\delta_{\mathcal{B}}$ is independent of the choice of symplectic basis $\mathcal{B}$.*

Proof. Take $\sigma \in \mathrm{O}(V)$ and $\tau \in \mathrm{Sp}(V)$. We saw in the proof above that $z_{\sigma\tau\mathcal{B}} = z_{\mathcal{B}} + \delta_{\mathcal{B}}(\sigma) + \delta_{\mathcal{B}}(\tau)$, but also $z_{\sigma(\tau\mathcal{B})} = z_{\tau\mathcal{B}} + \delta_{\tau\mathcal{B}}(\sigma) = z_{\mathcal{B}} + \delta_{\mathcal{B}}(\tau) + \delta_{\tau\mathcal{B}}(\sigma)$, so $\delta_{\tau\mathcal{B}}(\sigma) = \delta_{\mathcal{B}}(\sigma)$. □

Exercise

Suppose that $\{u_i, v_i\}$ is a symplectic basis for V, that $\sigma \in \mathrm{Sp}(V)$, and that $\sigma u_j = \sum_i (a_{ij}u_i + b_{ij}v_i)$, $\sigma v_j = \sum_i (c_{ij}u_i + d_{ij}v_i)$, all j. Use Corollary 13.9 and the fact that $z_{\sigma\mathcal{B}} = z_{\mathcal{B}} + \delta_{\mathcal{B}}(\sigma)$ to show that

1. if $j \neq k$ then
$$\textstyle\sum_i (a_{ji}c_{ki} + a_{ki}c_{ji}) = \sum_i (a_{ji}d_{ki} + b_{ki}c_{ji}) = \sum_i (b_{ji}d_{ki} + b_{ki}d_{ji}) = 0,$$
2. $\sum_i (a_{ji}d_{ji} + b_{ki}c_{ji}) = 1$, all j, k, and
3. $\delta_{\mathcal{B}}(\sigma) = \sum_{i,j} [a_{ji}c_{ji}Q(u_j) + b_{ji}c_{ji} + b_{ji}d_{ji}Q(v_j)]$.

Recall that if V is nondefective then each transvection in $\mathrm{Sp}(V)$ has the form $\tau = \tau_{u,a}$, with $u \neq 0$, $a \neq 0$, where $\tau v = v + aB(v,u)u$, all $v \in V$. In order for $\tau_{u,a}$ to be in $\mathrm{O}(V)$

$$\begin{aligned} Q(\tau v) &= Q(v + aB(v,u)u) \\ &= Q(v) + aB(v,u)^2 + a^2 B(v,u)^2 Q(u) = Q(v), \end{aligned}$$

and therefore $aB(v,u)^2[1 + aQ(u)] = 0$, all $v \in V$. Since V is nondefective $1 + aQ(u) = 0$, so $Q(u) \neq 0$ and $a = 1/Q(u)$. Thus τ is determined by a nonzero nonsingular vector u, and

$$\tau v = v + \frac{B(v,u)}{Q(u)} u$$

for all $v \in V$. We will write $\tau = \tau_u$. The converse is clear – given any nonzero nonsingular vector u then the formula above determines a transvection in $\mathrm{O}(V)$.

Exercise

Verify the following facts about transvections in $\mathrm{O}(V)$.

1. $\tau_u^2 = 1$.
2. If $\sigma \in \mathrm{O}(V)$ then $\sigma\tau_u\sigma^{-1} = \tau_{\sigma u}$.
3. If $a \in F^\star$ then $\tau_{au} = \tau_u$.
4. If $u_1 \perp u_2$ then $\tau_{u_1}\tau_{u_2} = \tau_{u_2}\tau_{u_1}$.

Let us compute $\delta(\tau_u)$ for a transvection $\tau_u \in \mathrm{O}(V)$. Choose a symplectic basis $\mathcal{B} = \{u_i, v_i\}$ with $u_1 = u$. Then $\tau_u u_1 = u_1$, $\tau_u v_1 = v_1 + u_1/Q(u_1)$, and $\tau_u u_i = u_i$, $\tau_u v_i = v_i$ for all $i \geq 2$. Thus

$$z_{\tau\mathcal{B}} = u_1\Big(\frac{u_1}{Q(u_1)} + v_1\Big) + \sum_{i=2}^{k} u_i v_i = \frac{u_1^2}{Q(u_1)} + \sum_{i=1}^{k} u_i v_i = 1 + z_{\mathcal{B}},$$

and therefore $\delta(\tau_u) = 1$.

In particular δ maps $\mathrm{O}(V)$ *onto* $\mathbb{F}_2$. We define the *special orthogonal group* $\mathrm{SO}(V)$ to be $\operatorname{Ker}\delta$, so $[\mathrm{O}(V)\colon \mathrm{SO}(V)] = 2$.

Recall that a hyperbolic pair (u, v) in V requires $Q(u) = Q(v) = 0$, a generally stronger requirement than for a symplectic hyperbolic pair. As usual a subspace W of V is called *hyperbolic* if W is an orthogonal direct sum of hyperbolic planes.

Proposition 14.6. *Suppose W_1 is a totally singular subspace of V and let $\{u_1, \dots , u_k\}$ be a basis for W_1. Then there is another totally singular subspace W_2, with basis $\{v_1, \dots , v_k\}$, such that $W_1 \cap W_2 = 0$, each $H_i = \langle u_i, v_i\rangle$ is a hyperbolic plane, and $W_1 \oplus W_2 = \textcircled{\perp}_{i=1}^{k} H_i$, so $W_1 \oplus W_2$ is hyperbolic.*

Proof. Induction on k. Define $f \in W_1^*$ via $f(u_1) = 1$, $f(u_i) = 0$ for $2 \le i \le k$. By Corollary 2.3 there exists a vector $v \in V$ such that $f(u_i) = B(u_i, v)$, $1 \le i \le k$. For any $a \in F$ we have $B(u_i, v + au_1) = B(u_i, v) = \delta_{i1}$. Since $Q(v + ua_1) = Q(v) + a$ we may set $a = Q(v)$ and set $v_1 = v + au_1$ so that v_1 is singular, (u_1, v_1) is a hyperbolic pair, and $v_1 \perp u_2, \dots , u_k$. Set $H_1 = \langle u_1, v_1\rangle$, a hyperbolic plane, so $V = H_1 \textcircled{\perp} H_1^{\perp}$. The proof is complete if $k = 1$.

Set $V_1 = H_1^{\perp}$ and set $U_1 = \langle u_2, \dots , u_k\rangle \subseteq V_1$. By induction there exists $U_2 = \langle v_2, \dots , v_k\rangle \subseteq V_1$, totally singular, such that $U_1 \cap U_2 = 0$, each (u_i, v_i) is a hyperbolic pair, and if $H_i = \langle u_i, v_i\rangle$ then $U_1 \textcircled{\perp} U_2 = H_2 \textcircled{\perp} \cdots \textcircled{\perp} H_k$. Set $W_2 = \langle v_1, v_2, \dots , v_k\rangle = \langle v_1\rangle \textcircled{\perp} U_2$. Then W_2 is totally singular. Also

$$W_1 + W_2 = \langle u_i, v_i \colon 1 \le i \le k\rangle = H_1 \textcircled{\perp} H_2 \textcircled{\perp} \cdots \textcircled{\perp} H_k,$$

so $W_1 \cap W_2 = 0$ and $W_1 + W_2 = W_1 \oplus W_2$ is hyperbolic. □

Corollary 14.7. *If W_1 is totally singular in V then there exist a totally singular subspace W_2 and a nondefective subspace W_3, with* $\dim W_2 = \dim W_1$ *and $V = (W_1 \oplus W_2) \textcircled{\perp} W_3$.*

Proof. Take W_2 as in the Proposition and set $W_3 = (W_1 \oplus W_2)^{\perp}$. □

Corollary 14.8. *Suppose that V has maximal Witt index $m = n/2$. Then V is hyperbolic and there exist maximal totally singular subspaces W_1 and W_2, with $V = W_1 \oplus W_2$, such that W_1 has a basis $\{u_1, \dots , u_m\}$, W_2 has a basis $\{v_1, \dots , v_m\}$, with each $H_i = \langle u_i, v_i\rangle$ a hyperbolic plane with hyperbolic pair (u_i, v_i), and $V = \textcircled{\perp}_{i=1}^{m} H_i$.*

Proposition 14.9. *Suppose that V has maximal Witt index $m = n/2$ and let $W_1 = \langle u_1, \dots , u_m\rangle$, $W_2 = \langle v_1, \dots , v_m\rangle$ be as in Corollary 14.8 above. Let* $\operatorname{Stab} W_1$ *denote the subgroup of* $\mathrm{O}(V)$ *stabilizing W_1 as a subspace. Then relative to the basis $\{u_1, \dots , u_m, v_1, \dots , v_m\}$ the elements of* $\operatorname{Stab} W_1$ *are* *represented by matrices of the form* $\begin{bmatrix} A & D \\ 0 & (A^t)^{-1} \end{bmatrix}$, *with* $AD^t + DA^t = 0$ *and* $A^{-1}D$ *symmetric and having zero diagonal.*

Proof. Take $\sigma \in \operatorname{Stab} W_1$ and say $\sigma u_j = \sum_i a_{ij}u_i$, $\sigma v_j = \sum_i (d_{ij}u_i + c_{ij}v_i)$. Set $A = [a_{ij}]$, $D = [d_{ij}]$, and $C = [c_{ij}]$, so σ is represented by $\begin{bmatrix} A & D \\ 0 & C \end{bmatrix}$. Since $\sigma \in \mathrm{O}(V)$ we have

$$\begin{aligned} \delta_{ij} &= B(u_i, v_j) = B(\sigma u_i, \sigma v_j) \\ &= \sum_{k,\ell} [a_{ki}d_{\ell j}B(u_k, u_\ell) + a_{ki}c_{\ell j}B(u_k, v_\ell)] \\ &= \sum_{k,\ell} a_{ki}c_{\ell j}\delta_{k\ell} = \sum_k a_{ki}c_{kj}. \end{aligned}$$

Thus $A^tC = I$, and $C = (A^t)^{-1}$. Also

$$\begin{aligned} 0 &= B(\sigma v_i, \sigma v_j) \\ &= B\Big(\sum_k (d_{ki}u_k + c_{ki}v_k), \sum_\ell (d_{\ell j}u_\ell + c_{\ell j}v_\ell)\Big) \\ &= \sum_{k,\ell} (d_{ki}c_{\ell j}\delta_{k\ell} + c_{ki}d_{\ell j}\delta_{k\ell}) \\ &= \sum_k (d_{ki}c_{kj} + c_{ki}d_{kj}), \end{aligned}$$

or $D^tC + C^tD = 0$, and so $D^t(A^t)^{-1} + A^{-1}D = 0$. Multiply on the left by A and on the right by A^t to obtain $AD^t + DA^t = 0$, and AD^t is symmetric. Note next that

$$\begin{aligned} 0 &= Q(v_i) = Q(\sigma v_i) \\ &= Q\Big(\sum_j (d_{ji}u_j + c_{ji}v_j)\Big) \\ &= \sum_j Q(d_{ji}u_j + c_{ji}v_j) = \sum_j d_{ji}c_{ji}, \end{aligned}$$

the iith entry of $D^t(A^{-1})^t = (A^{-1}D)^t$. Thus $A^{-1}D$ has 0 diagonal. Finally $(A^{-1}D)^t = D^t(A^{-1})^t = A^{-1}(AD^t)(A^{-1})^t$, so $A^{-1}D$ is symmetric, being conjugate to a symmetric matrix. □

Exercise

(Converse to Proposition 14.9) If $\sigma \colon V \to V$ is represented by the matrix $\begin{bmatrix} A & D \\ 0 & (A^t)^{-1} \end{bmatrix}$, with $AD^t + DA^t = 0$ and $A^{-1}D$ symmetric and having zero diagonal, show that $\sigma \in \mathrm{O}(V)$.

Corollary 14.10. *Maintain the hypotheses of Proposition 14.9. Suppose that* $\sigma|_{W_1} = 1|_{W_1}$. *Then* σ *is represented by* $\begin{bmatrix} I & D \\ 0 & I \end{bmatrix}$, *with* D *symmetric and having 0 diagonal. Thus* $\{\sigma \in \mathrm{O}(V) \colon \sigma|_{W_1} = 1_{W_1}\}$ *is an abelian group.*

Proof. Clearly $A = I$, and $D + D^t = 0$, so D is symmetric. Observe that

$$0 = Q(v_i) = Q(\sigma v_i) = Q(\sum_j d_{ji}u_j + v_i) = d_{ii}$$

for all i. □

Proposition 14.11. *Suppose that* $\sigma_1|_{W_1} = 1_{W_1}$ *and* $\sigma_2|_{W_1} = 1_{W_1}$ *as in Corollary 14.10, say with* σ_i *represented by* $\begin{bmatrix} I & D_i \\ 0 & I \end{bmatrix}$. *If* D_1 *and* D_2 *are of the same rank then* σ_1 *and* σ_2 *are conjugate in* $\mathrm{O}(V)$.

Proof. Define bilinear forms B_1, B_2 on W_1 via $\widehat{B_i} = D_i$ relative to the basis $\{u_1, \dots, u_m\}$. Since each D_i is symmetric with 0 diagonal the forms B_i are alternate and symmetric, and they determine symplectic geometries on W_1. Also $\mathrm{rank}_{B_i} W_1 = \mathrm{rank}\, D_i$, so there is an isometry from (W_1, B_1) to (W_1, B_2). Thus there exists an invertible matrix M such that $D_2 = M^t D_1 M$. Define a linear transformation $\tau \colon V \to V$ so that τ is represented by $\begin{bmatrix} M^t & 0 \\ 0 & M^{-1} \end{bmatrix}$. Then $\tau \in \mathrm{O}(V)$ by the exercise above, and $\tau\sigma_1\tau^{-1}$ is represented by

$$\begin{bmatrix} M^t & 0 \\ 0 & M^{-1} \end{bmatrix} \begin{bmatrix} I & D_1 \\ 0 & I \end{bmatrix} \begin{bmatrix} (M^t)^{-1} & 0 \\ 0 & M \end{bmatrix}$$

$$= \begin{bmatrix} I & M^t D_1 M \\ 0 & I \end{bmatrix} = \begin{bmatrix} I & D_2 \\ 0 & I \end{bmatrix},$$

which represents σ_2, so $\tau\sigma_1\tau^{-1} = \tau_2$. □

Write $\mathcal{T} = \mathcal{T}(V) = \langle \tau \colon \tau \text{ an orthogonal transvection} \rangle$, so $\mathcal{T} \lhd \mathrm{O}(V)$.

Proposition 14.12. *The subgroup* $\mathcal{T}$ *generated by transvections is transitive on the set of all maximal totally singular subspaces of* V.

Proof. Take W_1 and W_2 to be maximal totally singular subspaces, $W_1 \neq W_2$. We produce first $\tau_1 \in \mathcal{T}$ such that $\dim(W_1 \cap \tau_1 W_2) > \dim(W_1 \cap W_2)$. Since $W_1 + W_2$ is not totally singular there exists $u = w_1 + w_2$, $w_i \in W_i$, with $Q(u) \neq 0$. Note that

$$B(u, w_2) = B(w_1 + w_2, w_2) = B(w_1, w_2) = Q(w_1 + w_2) = Q(u) \neq 0.$$

Take $\tau_1 = \tau_u \in \mathcal{T}$. Then

$$\tau_1 w_2 = w_2 + \frac{B(w_2, u)}{Q(u)} u = w_2 + u = w_1,$$

so $w_1 \in W_1 \cap \tau_1 W_2$, but $w_1 \notin W_1 \cap W_2$. If $v \in W_1 \cap W_2$ then $B(v, u) = B(v, w_1 + w_2) = 0$, so $\tau_1 v = v \in W_1 \cap \tau_1 W_2$. Thus $W_1 \cap W_2 \subseteq W_1 \cap \tau_1 W_2$ but $W_1 \cap W_2 \neq W_1 \cap \tau_1 W_2$. If $\tau_1 W_2 = W_1$ we are finished. If not we may apply the above argument to W_1 and $\tau_1 W_2$ to produce $\tau_2 \in \mathcal{T}$ with $\dim W_1 \cap \tau_2 \tau_1 W_2 > W_1 \cap \tau_1 W_2$, etc. □

Recall from Proposition 6.2 that if $\sigma \in \mathrm{O}(V)$ then $\mathrm{Fix}(\sigma) = \mathrm{Ker}(\sigma - 1) = \mathrm{Im}(\sigma - 1)^\perp$.

Proposition 14.13. *Choose σ in $\mathrm{O}(V)$ and η in the coset $\sigma\mathcal{T} \in \mathrm{O}(V)/\mathcal{T}$ such that $\mathrm{Fix}(\eta)$ is maximal with respect to inclusion. Then $\mathrm{Im}(\eta + 1)$ is totally singular, and for any totally singular subspace $W \supseteq \mathrm{Im}(\eta + 1)$ we have $W^\perp \subseteq \mathrm{Fix}(\eta)$.*

Proof. Suppose that there exists $u = \eta v + v$ such that $Q(u) \neq 0$, i.e.

$$\begin{aligned} 0 &\neq Q(u) = Q(\eta v + v) \\ &= Q(\eta v) + B(\eta v, v) + Q(v) \\ &= B(\eta v, v) + 2Q(v) = B(\eta v, v) \\ &= B(\eta v + v, v) = B(u, v) \neq 0. \end{aligned}$$

Set $\tau = \tau_u \in \mathcal{T}$. Then

$$\tau v = v + \frac{B(v, u)}{Q(u)} u = v + u = \eta v,$$

so $v = \tau \eta v$ and $v \in \mathrm{Fix}(\tau\eta)$, but $v \notin \mathrm{Fix}(\eta)$ (since $\eta v = v + u \neq v$). Note, though, that $\mathrm{Fix}(\eta) = \mathrm{Im}(\eta + 1)^\perp \subseteq (\eta v + v)^\perp = u^\perp = \mathrm{Fix}(\tau)$, and consequently if $\eta x = x$ then $\tau \eta x = \tau x = x$, i.e. $\mathrm{Fix}(\eta) \subseteq \mathrm{Fix}(\tau\eta)$. Since $\tau\eta$ is in the same coset in $\mathrm{O}(V)/\mathcal{T}$ as η we have contradicted the maximality of $\mathrm{Fix}(\eta)$. Thus $\mathrm{Im}(\eta + 1)$ is totally singular. If $W \supseteq \mathrm{Im}(\eta + 1)$ then $\mathrm{Fix}(\eta) = \mathrm{Im}(\eta + 1)^\perp \supseteq W^\perp$. □

Recall that if U is a subspace of V then $\mathrm{O}_U(V)$ denotes the subgroup of $\mathrm{O}(V)$ consisting of all $\sigma \in \mathrm{O}(V)$ such that $\sigma|_U = 1_U$, so $\mathrm{O}_U(V) \cong \mathrm{O}(U)$.

Proposition 14.14. *If W is a maximal totally singular subspace of V then $\mathrm{O}(V) = \mathcal{T} \cdot \mathrm{O}_{W^\perp}(V)$.*

Proof. Take $\sigma \in \mathrm{O}(V)$. Choose η in the coset $\sigma\mathcal{T} \in \mathrm{O}(V)/\mathcal{T}$ with $\mathrm{Fix}(\eta)$ maximal relative to inclusion, and therefore $\mathrm{Im}(\eta + 1)$ is totally singular by Proposition 14.13. Choose W_σ to be a maximal totally singular subspace with $W_\sigma \supseteq \operatorname{Im}(\eta+1)$, so $W_\sigma^\perp \subseteq \operatorname{Fix}(\eta)$, also by Proposition 14.13. Thus $\eta \in \mathrm{O}_{W_\sigma^\perp}(V)$. By Proposition 14.12 there exists $\tau \in \mathcal{T}$ such that $\tau W = W_\sigma$, and hence $\tau W^\perp = W_\sigma^\perp$. Thus $\mathrm{O}_{W_\sigma^\perp} = \mathrm{O}_{\tau W^\perp} = \tau \, \mathrm{O}_{W^\perp} \tau^{-1}$, and hence $\mathcal{T}\,\mathrm{O}_{W_\sigma^\perp} = \mathcal{T}\tau\,\mathrm{O}_{W^\perp}\tau^{-1} = \mathcal{T}\,\mathrm{O}_{W^\perp}\tau^{-1} = \mathrm{O}_{W^\perp}\mathcal{T}\tau^{-1} = \mathcal{T}\,\mathrm{O}_{W^\perp}$. Thus

$$\sigma \in \mathcal{T}\eta \subseteq \mathcal{T}\,\mathrm{O}_{W_\sigma^\perp} = \mathcal{T}\,\mathrm{O}_{W^\perp},$$

and $\mathrm{O}(V) = \mathcal{T}\,\mathrm{O}_{W^\perp}(V)$. □

Write $\mathrm{O}'(V)$ for the derived group of $\mathrm{O}(V)$. Note that $\mathrm{O}'(V) \le \mathrm{SO}(V)$ since $\mathrm{O}(V)/\mathrm{SO}(V) \cong \mathbb{F}_2$ is abelian.

Theorem 14.15. $\mathrm{O}'(V) \le \mathcal{T}(V)$.

Proof. Choose W_1 to be a maximal totally singular subspace of V. Then choose W_2 also maximal totally singular and W_3 as in Corollary 14.7 so that $V = (W_1 \oplus W_2) \mathbin{\bot} W_3$. Clearly $W_1^\perp = W_1 \mathbin{\bot} W_3$. We have $\mathrm{O}(V) = \mathcal{T}\,\mathrm{O}_{W_1^\perp}(V)$ by Proposition 14.14, so that

$$\mathrm{O}(V)/\mathcal{T} = \mathcal{T}\cdot \mathrm{O}_{W_1^\perp}/\mathcal{T} \cong \mathrm{O}_{W_1^\perp}/(\mathcal{T} \cap \mathrm{O}_{W_1^\perp}).$$

Set $V_1 = W_1 \oplus W_2 \subseteq V$, so V_1 has maximal Witt index. It is easy to see that $\mathrm{O}_{W_1^\perp}(V) \cong \mathrm{O}_{W_1}(V_1)$. In fact if $\sigma \in \mathrm{O}_{W_1^\perp}(V)$ then $\sigma|_{V_1} \in \mathrm{O}(V_1)$ since $V_1 = W_3^\perp$, and $\sigma|_{W_1} = 1_{W_1}$, so $\sigma|_{V_1} \in \mathrm{O}_{W_1}(V_1)$. Conversely if $\tau \in \mathrm{O}_{W_1}(V_1)$ then τ extends trivially to $\sigma \in \mathrm{O}_{W_1^\perp}(V)$ via $\sigma|_{W_3} = 1_{W_3}$. But then $\mathrm{O}_{W_1^\perp}(V)$ is abelian by Corollary 14.10, hence $\mathrm{O}(V)/\mathcal{T}$ is abelian, and $\mathcal{T} \ge \mathrm{O}'(V)$. □

Theorem 14.16. *If $(n, m, |F|) \ne (4, 2, 2)$ then $\mathrm{O}(V)$ is generated by transvections, i.e. $\mathrm{O}(V) = \mathcal{T}(V)$.*

Proof. Fix $\sigma \in \mathrm{O}(V)$. Choose $\eta \in \sigma\mathcal{T}$ with $\operatorname{Fix}(\eta)$ maximal, and choose W_1 maximal totally singular with $W_1 \supseteq \operatorname{Im}(\eta+1)$, and therefore $W_1^\perp \subseteq \operatorname{Fix}(\eta)$. Choose W_2 and W_3 as in Corollary 14.7, so that $V = (W_1 \oplus W_2) \mathbin{\bot} W_3$, $W_1^\perp = W_1 \mathbin{\bot} W_3$, and $W_3^\perp = W_1 \oplus W_2$. Set $V_1 = W_1 \oplus W_2$, and write $\gamma = \eta|_{V_1}$.

Suppose first that $|F| > 2$. Then, by Corollary 14.10, γ is represented by a matrix of the form $\begin{bmatrix} I & D \\ 0 & I \end{bmatrix}$. Take $a \in F$, $a \ne 0, 1$, and let γ_1, $\gamma_2 \in \mathrm{O}_{W_1}(V_1)$ be represented by $\begin{bmatrix} I & aD \\ 0 & I \end{bmatrix}$, $\begin{bmatrix} I & (1+a)D \\ 0 & I \end{bmatrix}$, respectively. Then $\gamma_1\gamma_2 = \gamma$. Since D, aD, and $(1+a)D$ are of the same rank γ_1, γ_2, and γ are all conjugate in $\mathrm{O}(V_1)$. Extend γ, γ_1, and γ_2 (trivially on W_3) to η, η_1, $\eta_2 \in \mathrm{O}_{W_1^\perp}$, respectively, with $\eta = \eta_1\eta_2$. The conjugating elements may also be extended (trivially on W_3), so we may assume that $\eta^{\tau_1} = \eta_1$, $\eta^{\tau_2} = \eta_2$, $\tau_i \in \mathrm{O}(V)$. Since $\mathrm{O}'(V) \le \mathcal{T}$ we have $\mathrm{O}(V)/\mathcal{T}$ abelian, and hence

$\eta\mathcal{T} = \eta_1\eta_2\mathcal{T} = \eta_1^\tau\eta_2^\tau\mathcal{T} = \eta^2\mathcal{T}$, and therefore $\mathcal{T} = \eta\mathcal{T}$, or $\eta \in \mathcal{T}$. Thus $\sigma \in \eta\mathcal{T} = \mathcal{T}$, and $\mathrm{O}(V) = \mathcal{T}$.

Suppose next that $|F| = 2$. If $m = 0$ then $\eta = 1$ and $\sigma \in \eta\mathcal{T} = \mathcal{T}$. If $m = 1$ then γ is represented by $\begin{bmatrix} 1 & d \\ 0 & 1 \end{bmatrix}$, and $\mathrm{diag}(d) = d = 0$, so $\gamma = 1$, $\eta = 1$, and $\sigma \in \mathcal{T}$. Assume then that $m \geq 2$, and hence $n > 4$. Again γ is represented by $\begin{bmatrix} I & D \\ 0 & I \end{bmatrix}$, D symmetric with zero diagonal. As in the proof of Proposition 14.11 we may view D as the representing matrix for an alternating bilinear form on W_1, determining a symplectic geometry on W_1. By Theorem 2.10 there is an invertible matrix M such that M^tDM is block diagonal of the form $\mathrm{diag}(S, \dots, S, 0)$, $S = \begin{bmatrix} 0 & 1 \\ 1 & 0 \end{bmatrix}$. The matrix $\begin{bmatrix} M^t & 0 \\ 0 & M^{-1} \end{bmatrix}$ represents an element of $\mathrm{O}(V_1)$ (which extends trivially to an element of $\mathrm{O}(V)$), and

$$\begin{bmatrix} M^t & 0 \\ 0 & M^{-1} \end{bmatrix} \begin{bmatrix} I & D \\ 0 & I \end{bmatrix} \begin{bmatrix} M^t & 0 \\ 0 & M^{-1} \end{bmatrix}^{-1} = \begin{bmatrix} I & M^tDM \\ 0 & I \end{bmatrix}.$$

As above we want to show that $\eta \in \mathcal{T}$, and hence $\sigma \in \mathcal{T}$. Since

$$\begin{bmatrix} I & D_1 \\ 0 & I \end{bmatrix} \begin{bmatrix} I & D_2 \\ 0 & I \end{bmatrix} = \begin{bmatrix} I & D_1 + D_2 \\ 0 & I \end{bmatrix}$$

it will be sufficient to assume that D has exactly one S on the diagonal. Thus we are assuming that $\eta u_i = u_i$, all i, $\eta v_i = v_i$ if $i > 2$, $\eta v_1 = u_2 + v_1$, $\eta v_2 = u_1 + v_2$, $\eta w = w$ if $w \in V_3$. Write $V = H_1 \oplus H_2 \oplus W_4$, $H_i = \langle u_i, v_i \rangle$ and $W_4 = (H_1 + H_2)^\perp$, and note that $\eta|_{W_4} = 1_{W_4}$, $W_4 \neq 0$ ($n > 4$). Since W_4 is nondefective there exists $w \in W_4$ such that $Q(w) = 1$. Then also

$$Q(w + u_1) = Q(w + u_2) = Q(w + u_1 + u_2) = Q(w) = 1,$$

so there are transvections $\tau_1 = \tau_{w+u_1}$, $\tau_2 = \tau_{w+u_2}$, $\tau_3 = \tau_{w+u_1+u_2}$, and $\tau_4 = \tau_w$ in $\mathcal{T}$. A mildly tedious calculation shows that $\eta = \tau_4\tau_3\tau_2\tau_1 \in \mathcal{T}$, so $\sigma \in \eta\mathcal{T} = \mathcal{T}$, and $\mathrm{O}(V) = \mathcal{T}$. □

Recall Proposition 8.15: there is a unique algebra anti-automorphism α of the Clifford algebra $C(V)$ such that $\alpha|_V = 1_V$.

Proposition 14.17. *Suppose that $w_1, \dots, w_k$ are nonzero nonsingular vectors in V and set $\tau_i = \tau_{w_i}$, $1 \leq i \leq k$. If $\tau_1\tau_2\cdots\tau_k = 1$ then k is even and $\prod_{i=1}^k Q(w_i) \in F^{\times 2}$.*

Proof. Apply the pseudodeterminant:

$$\delta(\tau_1\tau_2\cdots\tau_k) = \sum_{i=1}^k \delta(\tau_i) = k \cdot 1 = \delta(1) = 0,$$

so k is even. For any $v \in V$ we have

$$v = \tau_1 \cdots \tau_k v = w_1 \cdots w_k v w_k^{-1} \cdots w_1^{-1} = (w_1 \cdots w_k) v (w_1 \cdots w_k)^{-1}$$

by Proposition 13.4, and so $w_1 \cdots w_k \in Z(C(V))$. In fact, since k is even, $w_1 \cdots w_k \in Z(C_0(V))$, and thus $w_1 \cdots w_k = a + bz$, $a, b \in F$, by Theorem 13.12. Since $z \notin Z(C(V))$ (e.g. $zu_i = u_i z + u_i$), it follows that $b = 0$ and so $w_1 \cdots w_k \in F^\star$.

The remainder of the proof proceeds exactly as in the proof of Proposition 9.1, using the anti-automorphism α of $C(V)$. □

For any $\sigma \in \mathcal{T}$ write $\sigma = \tau_{w_1} \tau_{w_2} \cdots \tau_{w_k}$, a product of transvections. Just as in Chapter 9 we have a well-defined homomorphism $\theta \colon \mathcal{T} \to F^\star / F^{\times 2}$, the *spinor norm*, defined via $\theta(\sigma) = \prod_{i=1}^{k} Q(w_i) F^{\times 2}$.

Assume for most of the remainder of the chapter that $(n, m, |F|) \neq (4, 2, 2)$, and hence that $\mathrm{O}(V) = \mathcal{T}$, i.e. $\mathrm{O}(V)$ is generated by transvections.

Proposition 14.18. *If V is singular ($m > 0$) then the spinor norm θ maps $\mathrm{SO}(V)$ onto $F^\star / F^{\times 2}$.*

Proof. Let (u, v) be a hyperbolic pair in V. If $a \in F^\star$ set $w_1 = u + av$, $w_2 = u + v$, and $\sigma = \tau_{w_1} \tau_{w_2} \in \mathrm{SO}(V)$. Then $\theta(\sigma) = Q(w_1) Q(w_2) F^{\times 2} = a \cdot 1 F^{\times 2}$. □

The proofs of the next proposition and its corollary are exactly the same as the proofs of Proposition 9.3 and Corollary 9.4.

Proposition 14.19. *Suppose that w_1, w_2 are nonzero nonsingular vectors with $Q(w_1) = Q(w_2)$. Then $\tau_{w_1} \tau_{w_2} \in \mathrm{O}'(V)$.*

Corollary 14.20. *Suppose that $w_1, \ldots, w_k, z_1, \ldots, z_k \in V$, and that*

$$Q(w_i) = Q(z_i) \neq 0,$$

all i. Then $\tau_{w_1} \cdots \tau_{w_k}$ and $\tau_{z_1} \cdots \tau_{z_k}$ lie in the same coset of $\mathrm{O}'(V)$ in $\mathrm{O}(V)$.

Exercise

If V is a hyperbolic plane with hyperbolic pair (u, v) show that relative to $\{u, v\}$ any transvection of V is represented by a matrix of the form $\begin{bmatrix} 0 & b^{-1} \\ b & 0 \end{bmatrix}$, $b \in F^\star$, and that

$$\mathrm{SO}(2, F) = \left\{ \begin{bmatrix} a & 0 \\ 0 & a^{-1} \end{bmatrix} : a \in F^\star \right\} \cong F^\star.$$

Compute the spinor norm θ. Show that $\operatorname{Ker} \theta = \mathrm{O}(2, F^\square)$ and that every element of $\mathrm{O}(V) \setminus \mathrm{SO}(V)$ is a transvection.

Proposition 14.21. *Suppose that* $V = V_1 \oplus V_2$. *If* $\sigma_i \in \mathrm{O}(V_i)$ *and* $\sigma = (\sigma_1 \oplus 1_{V_2})(1_{V_1} \oplus \sigma_2) \in \mathrm{O}(V)$ *then* $\theta_V(\sigma) = \theta_{V_1}(\sigma_1)\theta_{V_2}(\sigma_2)$.

Proof. Write each σ_i as a product of transvections of V_i. Extend all the transvections to transvections of V by the same formulas. □

Proposition 14.22. *Suppose that* W *is a nondefective subspace of* V, *so that* $V = W \oplus W^\perp$, *and suppose that* $Q(W) = Q(V)$. *If* $\mathrm{SO}(W) \cap \operatorname{Ker}\theta_W = \mathrm{O}'(W)$ *then also* $\mathrm{SO}(V) \cap \operatorname{Ker}\theta_V = \mathrm{O}'(V)$.

Proof. Since $\mathrm{O}(V)/\mathrm{SO}(V)$ and $\mathrm{O}(V)/\operatorname{Ker}\theta_V$ are both abelian we have in general that $\mathrm{O}'(V) \le \mathrm{SO}(V) \cap \operatorname{Ker}\theta_V$. Suppose that $\sigma \in \mathrm{SO}(V) \cap \operatorname{Ker}\theta_V$ and write $\sigma = \tau_1 \cdots \tau_k$, a product of k (which is even) transvections, say with $\tau_i = \tau_{z_i}$, each i. Choose $w_1, \dots, w_k \in W$ such that $Q(w_i) = Q(z_i)$, all i, and set $\eta = \tau_{w_1} \cdots \tau_{w_k} \in \mathrm{SO}(V)$. Then $\theta_V(\eta) = \theta_V(\sigma) = 1 \pmod{F^{\times 2}}$, so $\sigma \in \eta\mathrm{O}'(V)$ by Corollary 14.20. Note though that $\eta|_W \in \mathrm{SO}(W)$ and $\theta_W(\eta|_W) = \theta_V(\eta) = 1 \pmod{F^{\times 2}}$, so $\eta|_W \in \mathrm{SO}(W) \cap \operatorname{Ker}\theta_W = \mathrm{O}'(W)$. Clearly then $\eta \in \mathrm{O}'(V)$, as it is extended trivially on $W^\perp$ from $\eta|_W$, and so $\sigma \in \mathrm{O}'(V)$. □

Proposition 14.23. *If* V *is singular* ($m > 0$) *then*

$$\mathrm{O}'(V) = \mathrm{SO}(V) \cap \operatorname{Ker}\theta_V.$$

Proof. Let W be a hyperbolic plane in V, so $V = W \oplus W^\perp$. Then $Q(W) = F$ (see the proof of Proposition 14.18), so $Q(V) = Q(W) = F$. By the exercise above we see that $\mathrm{SO}(W) \cap \operatorname{Ker}\theta_W$ consists of all σ represented by matrices of the form $\begin{bmatrix} a^2 & 0 \\ 0 & a^{-2} \end{bmatrix}$, $a \in F^\star$. Since

$$\begin{bmatrix} a & 0 \\ 0 & a^{-1} \end{bmatrix}\begin{bmatrix} 0 & 1 \\ 1 & 0 \end{bmatrix}\begin{bmatrix} a^{-1} & 0 \\ 0 & a \end{bmatrix}\begin{bmatrix} 0 & 1 \\ 1 & 0 \end{bmatrix} = \begin{bmatrix} a^2 & 0 \\ 0 & a^{-2} \end{bmatrix}$$

we conclude that $\mathrm{O}'(W) = \mathrm{SO}(W) \cap \operatorname{Ker}\theta_W$. Thus $\mathrm{O}'(V) = \mathrm{SO}(V) \cap \operatorname{Ker}\theta_V$ by Proposition 14.22. □

Proposition 14.24. *If* $\dim V = n \ge 4$ *then* $\mathrm{O}'(V) = \mathrm{SO}'(V)$.

Proof. Since $\mathrm{O}(V) = \mathcal{T}(V)$ is generated by transvections it will suffice to show that every commutator of two transvections is in $\mathrm{SO}'(V)$. Take w, z nonzero and nonsingular, and set $W = \langle w, z\rangle$. Then $W^\perp$ is not totally singular, or else $W^\perp \subseteq W^{\perp\perp} = W$, and then $\dim W^\perp = n - \dim W \ge 2$. But then $W = W^\perp$ is totally singular, contradicting $Q(w) \ne 0$. Choose $x \in W^\perp$ with $Q(x) \ne 0$. Then τ_x commutes with both τ_w and τ_z, and so

$$[\tau_w, \tau_z] = [\tau_w\tau_x, \tau_z\tau_x] \in \mathrm{SO}'(V).$$

□

Assume until further notice that $\dim V = n \geq 4$ *and that* $m > 0$*, i.e.* V *is singular. Continue to assume that* $(n, m, |F|) \neq (4, 2, 2)$.

Choose a singular vector $x \in V$ and choose $y \in x^{\perp}$. Then define $\eta_{x,y}\colon x^{\perp} \to x^{\perp}$ by setting

$$\eta_{x,y}(w) = w + B(w, y)x,$$

all $w \in x^{\perp}$. As in Chapter 6 (see page 52) it is easily verified that

$$\eta_{x,y} \in \mathrm{GL}(x^{\perp}), \quad \eta_{x,0} = 1_{x^{\perp}}, \quad \text{and} \quad \eta_{x,y+z} = \eta_{x,y}\eta_{x,z},$$

so $\eta_{x,y}^{-1} = \eta_{x,-y}$. If $w \in x^{\perp}$ then $Q(\eta_{x,y}w) = Q(w + B(w, y)x) = Q(w)$, since $w \perp x$ and $Q(x) = 0$, so $\eta_{x,y} \in \mathrm{O}(x^{\perp})$.

The next proposition and its corollaries are proved exactly as Proposition 6.16 and its corollaries were proved.

Proposition 14.25. *If* $x \in V$ *is singular and* $y \in x^{\perp}$ *then the isometry* $\eta_{x,y}$ *of* $x^{\perp}$ *extends uniquely to an isometry* $\rho_{x,y} \in \mathrm{O}(V)$.

Corollary 14.26. *If* y, $z \in x^{\perp}$ *then* $\rho_{x,y+z} = \rho_{x,y}\rho_{x,z}$.

Corollary 14.27. *If* $a \in F^{\star}$ *then* $\rho_{ax,y} = \rho_{x,ay}$.

Corollary 14.28. $\rho_{x,y}^{-1} = \rho_{x,-y}$.

Corollary 14.29. $\rho_{x,y} = 1_V$ *if and only if* $y \in \langle x \rangle = \mathrm{rad}\, x^{\perp}$.

Corollary 14.30. *If* $\tau \in \mathrm{O}(V)$ *then* $\tau\rho_{x,y}\tau^{-1} = \rho_{\tau x,\tau y}$.

Exercise

Show that the extension of $\eta_{x,y}$ to $\rho_{x,y}$ can be described explicitly as follows. Choose $v \in V$ so that (x, v) is a hyperbolic pair, and set $H = \langle x, v \rangle$, so $V = H \oplus H^{\perp}$. Set $y_0 = y + B(y, v)x$. Show that $y_0 \in H^{\perp}$, that

$$\rho_{x,y}v = v + y_0 + Q(y_0)\, x,$$

and also that

$$\rho_{x,y}w = w + B(w, y)x = w + B(w, y_0)x$$

for all $w \in H^{\perp}$.

The orthogonal transformations $\rho_{x,y}$ defined above are called *pseudo-transvections* of V. For each singular vector x set $K_x = \{\rho_{x,y}\colon y \in x^{\perp}\}$, an abelian subgroup of $\mathrm{Stab}_{\mathrm{O}(V)}(x) \leq \mathrm{O}(V)$. Define

$$\Omega(V) = \langle \cup K_x\colon x \text{ singular in } V \rangle = \langle \rho_{x,y}\colon x \text{ singular},\ y \in x^{\perp} \rangle \leq \mathrm{O}(V).$$

Note that $\Omega(V) \triangleleft \mathrm{O}(V)$ by Corollary 14.30.

Proposition 14.31. $\Omega(V) \leq \mathrm{O}'(V)$.

Proof. Take a pseudotransvection $\rho_{x,y} \in \Omega$ with $y \neq 0$. Every $y \neq 0$ is a sum of nonsingular vectors by Proposition 12.8, and $\rho_{x,y+z} = \rho_{x,y}\rho_{x,z}$, so it will be sufficient to assume that $Q(y) \neq 0$ and to show that $\rho_{x,y} \in \mathrm{O}'(V)$. Note that $Q(y + Q(y)x) = Q(y) + Q(y)B(y,x) + Q(y)^2Q(x) = Q(y)$ so by the Witt Extension Theorem (12.10) there exists $\sigma \in \mathrm{O}(V)$ such that $\sigma y = y + Q(y)x$. Let τ_1 and τ_2 be the transvections τ_y and $\tau_{y+Q(y)x}$, respectively. Then $\sigma\tau_1\sigma^{-1} = \tau_{\sigma y} = \tau_2$, so $\tau_2\tau_1 = \sigma\tau_1\sigma^{-1}\tau_1^{-1} \in \mathrm{O}'(V)$, and it will suffice to show that $\tau_2\tau_1 = \rho_{x,y}$.

Take $z \in x^\perp$. Then

$$\begin{aligned}
\tau_2\tau_1 z &= \tau_2\big(z + \frac{B(z,y)}{Q(y)}y\big) \\
&= z + \frac{B(z,y)}{Q(y)}y + \frac{B(z + Q(z)y/Q(y), y + Q(y)x)}{Q(y)}(y + Q(y)x) \\
&= z + \frac{B(z,y)}{Q(y)}y + \frac{B(z,y)}{Q(y)}(y + Q(y)x) \\
&= z + B(z,y)x = \eta_{x,y}z,
\end{aligned}$$

so $\rho_{x,y} = \tau_2, \tau_1$ by Proposition 14.25. □

As in Chapter 6 we define

$$\mathcal{C} = \{x \in V \colon Q(x) = 0\},$$

the *quadric cone*, and

$$\mathbb{P}\mathcal{C} = \{[x] \in \mathbb{P}_{n-1}(V) \colon x \neq 0 \text{ and } x \in \mathcal{C}\},$$

the *projective quadric cone.*

Proposition 14.32. $\mathrm{O}(V)$ *acts faithfully on* $\mathbb{P}\mathcal{C}$.

Proof. Let K be the kernel of the action and take $\tau \in K$. Let (u, v) be a hyperbolic pair in V, set $H = \langle u, v\rangle$, and choose w nonzero and nonsingular in $H^\perp$. Set $x = u + Q(w)v + w$ and note that $Q(x) = 2Q(w) = 0$, so $[x] \in \mathbb{P}\mathcal{C}$. Say $\tau u = au$, $\tau v = bv$, and $\tau x = cx$. Then

$$cx = cu + cQ(w)v + cw = \tau x = au + bQ(w)v + \tau w.$$

Since $\tau H = H$ and $\tau H^\perp = H^\perp$ we may conclude that $\tau w = cw$ and $a = b = c$. By Proposition 12.8 $H^\perp$ has a basis of vectors such as w, so $\tau = c \cdot 1_V$. In particular $Q(w) = Q(\tau w) = Q(cw) = c^2Q(w)$, so $c^2 = 1$, $c = 1$, and hence $\tau = 1_V$. □

For any $x \neq 0$ in V write $T_x = \mathcal{C} \cap x^\perp$, the *trace* of $\mathcal{C}$ in $x^\perp$, and set $\mathbb{P}T_x = \{[v] \in \mathbb{P}V \colon 0 \neq v \in T_x\}$, the *projective trace.*

Proposition 14.33. *If x is singular in V then K_x acts transitively on $\mathbb{P}C \setminus \mathbb{P}T_x$.*

Proof. Take $[u] \neq [v]$ in $\mathbb{P}C \setminus \mathbb{P}T_x$. We may assume that $B(x, u) = B(x, v) = 1$, so (x, u) is a hyperbolic pair. Set $H = \langle x, u \rangle$, so $V = H \oplus H^\perp$. Write $v = ax + bu + z$, a, $b \in F$, $z \in H^\perp$. Then $1 = B(x, v) = B(x, ax + bu + z) = b$ and $0 = Q(v) = Q(ax + u + z) = Q(ax + u) + Q(z) = a + Q(z)$. Thus $v = u + Q(z)x + z = \rho_{x,z} u$ by the exercise on page 139. Thus $\rho_{x,z}[u] = [v]$. □

Proposition 14.34. *If $[x]$, $[y] \in \mathbb{P}C$ then $\mathbb{P}C \neq \mathbb{P}T_x \cup \mathbb{P}T_y$.*

Proof. Suppose first that $x \perp y$. Since $\dim V \geq 4$ $y^\perp$ is not totally singular, there exists $u \in y^\perp$ such that (x, u) is a hyperbolic pair. Set $H = \langle x, u \rangle$, so $V = H \oplus H^\perp$ and $y \in H^\perp$. Choose $v \in H^\perp$ so that (y, v) is a hyperbolic pair, and set $z = u + v$. Then $Q(z) = Q(u) + Q(v) = 0$, so $[z] \in \mathbb{P}C$ and $B(x, z) = B(x, u + v) = 1 = B(y, z)$, so $[z] \notin \mathbb{P}T_x \cup \mathbb{P}T_y$.

Suppose next that $B(x, y) \neq 0$, so we may assume that $B(x, y) = 1$. Set $W = \langle x, y \rangle$, so $V = W \oplus W^\perp$. Choose $w \in W^\perp$ with $Q(w) \neq 0$ and set $z = Q(w)x + y + w$. Then $Q(z) = Q(Q(w)x + y) + Q(w) = 2Q(w) = 0$, so $[z] \in \mathbb{P}C$, and $B(x, z) = B(x, Q(w)x + y + w) = 1$, $B(y, z) = Q(w) \neq 0$, and $[z] \notin \mathbb{P}T_x \cup \mathbb{P}T_y$. □

Proposition 14.35. *If $m(V) \geq 2$ and $[x] \neq [y]$ in $\mathbb{P}C$ then $\mathbb{P}T_x \not\subseteq \mathbb{P}T_y$.*

Proof. If $x \perp y$ then as in the proof above there exist u, $v \in V$ such that (x, u) and (y, v) are hyperbolic pairs and $\langle x, u \rangle \perp \langle y, v \rangle$. Then $B(x, v) = 0$, $B(y, v) = 1$, and $[v] \in \mathbb{P}T_x \setminus \mathbb{P}T_y$.

Suppose then that $B(x, y) \neq 0$, so we may assume that $B(x, y) = 1$. Set $W = \langle x, y \rangle$, so $V = W \oplus W^\perp$. Since $m \geq 2$ there exists a singular vector $w \in W^\perp$. Set $z = x + w$. Then $Q(z) = Q(x) + Q(w) = 0$, so $z \in \mathbb{P}C$, and $B(x, z) = B(x, x + w) = 0$, $B(y, z) = B(y, x + w) = 1$, and $[z] \in \mathbb{P}T_x \setminus \mathbb{P}T_y$. □

Proposition 14.36. *$\Omega(V)$ acts transitively on $\mathbb{P}C$.*

Proof. If $[x] \neq [y]$ in $\mathbb{P}C$ choose $[z] \in \mathbb{P}C \setminus (\mathbb{P}T_x \cup \mathbb{P}T_y)$, hence $[x]$, $[y] \in \mathbb{P}C \setminus \mathbb{P}T_z$. By Proposition 14.33 there exists $\sigma \in K_z \leq \Omega$ with $\sigma[x] = [y]$. □

Proposition 14.37. *If $m = 1$ then $\Omega(V)$ is doubly transitive on $\mathbb{P}C$.*

Proof. If $[x] \in \mathbb{P}C$ then $T_x = C \cap x^\perp = \langle x \rangle$, so $\mathbb{P}T_x = \{[x]\}$. But now $K_x \subseteq \mathrm{Stab}_\Omega([x])$, and K_x is transitive on $\mathbb{P}C \setminus \mathbb{P}T_x = \mathbb{P}C \setminus \{[x]\}$ by Proposition 14.33, so Ω is doubly transitive. □

Proposition 14.38. *If $(n, m) \neq (4, 2)$ and $[x] \in \mathbb{P}C$ then $\Omega(V)$ is doubly transitive on $\mathbb{P}T_x$.*

Proof. If $m = 1$ this is immediate from Proposition 14.37, so assume that $m \geq 2$ and hence $n \geq 6$. Choose $y \in V$ such that (x, y) is a hyperbolic pair and set $H = \langle x, y \rangle$, so $V = H \oplus H^{\perp}$. Observe then that

$$\mathbb{P}T_x = \{[u] \colon u = ax + w,\ a \in F,\ w \in H^{\perp},\ Q(w) = 0\}.$$

Write $\Omega_H = \Omega_H(V)$ for the pointwise stabilizer of H in $\Omega(V)$; clearly $\Omega_H \cong \Omega(H^{\perp})$. By Proposition 14.36 Ω_H is transitive on $\mathbb{P}\mathcal{C}(H^{\perp})$. Take $u = ax + w$ and $v = bx + z$ so that $[u], [v] \in \mathbb{P}T_x \setminus \{[x]\}$, with $[u] \neq [v]$. Take $\sigma \in \Omega_H$ such that $\sigma[w] = [z]$. Since $\sigma x = x$ we have $\sigma[u] = \sigma[ax + w] = [cx + z]$ for some $c \in F$. Choose $y_1 \in H^{\perp}$ such that $B(z, y_1) = b + c$. Then $\rho_{x,y_1}\sigma \in \Omega$, and

$$\begin{aligned} \rho_{x,y_1}\sigma[u] &= \rho_{x,y_1}[cx + z] = [cx + z + B(z, y_1)x] \\ &= [cx + z + (b + c)x] = [bx + z] = [v]. \end{aligned}$$

□

Remark

If $m \geq 2$ (but $(n, m) \neq (4, 2)$) then in fact Ω has rank 3 on $\mathbb{P}\mathcal{C}$. Specifically, the orbits of $\mathrm{Stab}_\Omega([x])$ are $\{[x]\}$, $\mathbb{P}T_x \setminus \{[x]\}$, and $\mathbb{P}\mathcal{C} \setminus \mathbb{P}T_x$ for each $[x] \in \mathbb{P}\mathcal{C}$.

Proposition 14.39. *If $(n, m) \neq (4, 2)$ then $\Omega(V)$ acts primitively on $\mathbb{P}\mathcal{C}$.*

Proof. If $m = 1$ this follows immediately from Proposition 14.37, so assume that $m \geq 2$ and therefore $n \geq 6$. Suppose that $\mathcal{B} \subseteq \mathbb{P}\mathcal{C}$ with $|\mathcal{B}| \geq 2$ and either $\sigma\mathcal{B} = \mathcal{B}$ or $\sigma\mathcal{B} \cap \mathcal{B} = \emptyset$ for all $\sigma \in \Omega$. Choose $[v] \neq [w]$ in $\mathcal{B}$. We show first that we may assume that $B(v, w) \neq 0$. If $B(v, w) = 0$ than $[w] \in \mathbb{P}T_v \setminus \{[v]\}$ and $\mathrm{Stab}_\Omega([v])$ is transitive on $\mathbb{P}T_v \setminus \{[v]\}$ by Proposition 14.38. If $\sigma \in \mathrm{Stab}_\Omega([v])$ then $[v] \in \mathcal{B} \cap \sigma\mathcal{B}$, so $\sigma\mathcal{B} = \mathcal{B}$. Thus $\mathbb{P}T_v \setminus \{[v]\} = \{\sigma[w] \colon \sigma \in \mathrm{Stab}([v])\} \subseteq \mathcal{B}$, and so $\mathbb{P}T_v \subseteq \mathcal{B}$. Likewise $\mathbb{P}T_w \subseteq \mathcal{B}$. Choose $[z] \in \mathbb{P}\mathcal{C}$ such that $[z] \in \mathbb{P}T_V \setminus \mathbb{P}T_w$ and $[z] \neq [v]$ (Proposition 14.35). Then we have $[z] \in \mathbb{P}T_v \subseteq \mathcal{B}$, but $B(z, w) \neq 0$ since $[z] \notin \mathbb{P}T_w$, and we may replace $[v]$ by $[z]$.

Simply assume, then, that $B(v, w) \neq 0$, and hence $[w] \in \mathbb{P}\mathcal{C} \setminus \mathbb{P}T_v$. By Proposition 14.33 we have

$$\mathbb{P}\mathcal{C} \setminus \mathbb{P}T_v = \{\sigma[w] \colon \sigma \in K_v\} \subseteq \cup\{\sigma\mathcal{B} \colon \sigma \in K_v\} \subseteq \mathcal{B}$$

(as above), and likewise $\mathbb{P}\mathcal{C} \setminus \mathbb{P}T_w \subseteq \mathcal{B}$. Choose $[x] \in \mathbb{P}T_v \setminus \mathbb{P}T_w$, $[x] \neq [v]$. As above $\mathbb{P}T_v \setminus \{[v]\} = \{\sigma[x] \colon \sigma \in \mathrm{Stab}([v])\} \subseteq \mathcal{B}$, so $\mathbb{P}T_v \subseteq \mathcal{B}$. But now we have $\mathcal{B} \supseteq \mathbb{P}T_v \cup (\mathbb{P}\mathcal{C} \setminus \mathbb{P}T_v) = \mathbb{P}\mathcal{C}$, so $\mathcal{B} = \mathbb{P}\mathcal{C}$ and Ω is primitive. □

Proposition 14.40. *Suppose that H is a hyperbolic plane in V. Then*

$$\mathrm{O}(V) = \Omega(H) \cdot \mathrm{O}_{H^{\perp}}(V).$$

Proof. It will suffice to show that every transvection τ_x is in $\Omega\,\mathrm{O}_{H^\perp}$. Take $x \in V$ with $Q(x) \neq 0$. Let (u, v) be a hyperbolic pair in H and set $y = Q(x)u + v$. Then $Q(y) = Q(x)$ so by the Witt Extension Theorem (12.10) there exists $\sigma \in \mathrm{O}(V)$ with $\sigma y = x$. Set $u_1 = \sigma u$ and $v_1 = \sigma v$. By Proposition 14.36 there exists $\eta \in \Omega$ such that $\eta[u] = [u_1]$. Since $B(u_1, v_1) = 1$ and

$$B(\eta v, u_1) = B(v, \eta^{-1}u_1) = B(v, au) = a$$

for some $a \in F^\star$ we see that $[v_1]$, $[\eta v] \in \mathbb{P}\mathcal{C}\backslash\mathbb{P}T_{u_1}$. Thus by Proposition 14.33 there exists $\gamma \in K_{u_1} \leq \Omega$ such that $\gamma\eta[v] = [v_1]$. Also $\gamma\eta[u] = \gamma[u_1] = [u_1]$ since $K_{u_1} \leq \mathrm{Stab}([u_1])$. Note now that $x = \sigma y \in \sigma H = \langle u_1, v_1\rangle = \gamma\eta\langle u, v\rangle = \gamma\eta H$. Set $z = (\gamma\eta)^{-1}x \in H$. Then

$$\tau_x = \tau_{\gamma\eta z} = (\gamma\eta)\tau_z(\gamma\eta)^{-1} = \tau_z(\tau_z(\gamma\eta)\tau_z^{-1}(\gamma\eta)^{-1}),$$

with $\tau_z \in \mathrm{O}_{H^\perp}$ and $\tau_z(\gamma\eta)\tau_z^{-1}(\gamma\eta)^{-1} \in \Omega$ since $\gamma\eta \in \Omega$ and $\Omega \lhd \mathrm{O}(V)$. □

Proposition 14.41. $\Omega(V) = \mathrm{O}'(V) = \mathrm{SO}'(V)$.

Proof. We know that $\Omega \leq \mathrm{O}'(V) \leq \mathrm{SO}(V)$ by Propositions 14.23 and 14.31, and that $\mathrm{O}'(V) = \mathrm{SO}'(V)$ by Proposition 14.24. Since $\Omega \leq \mathrm{SO}(V)$ it follows from Proposition 14.40 that $\mathrm{SO}(V) = \Omega\cdot\mathrm{SO}_{H^\perp}(V)$ for any hyperbolic plane H. Thus

$$\mathrm{SO}(V)/\Omega = \Omega\cdot\mathrm{SO}_{H^\perp}(V)/\Omega \cong \mathrm{SO}_{H^\perp}/(\Omega \cap \mathrm{SO}_{H^\perp}).$$

Since $\mathrm{SO}_{H^\perp} \cong \mathrm{SO}(H)$ is abelian (by the exercise on page 137) we conclude that $\mathrm{SO}'(V) \leq \Omega \leq \mathrm{O}'(V) = \mathrm{SO}'(V)$, and all are equal. □

Proposition 14.42. $\mathrm{O}''(V) = \mathrm{O}'(V)$.

Proof. Suppose first that $|F| \neq 2$ and choose $c \in F \setminus \{0, 1\}$. Choose a hyperbolic pair (u, v) in V and set $H = \langle u, v\rangle$. By the exercise on page 137 there exists $\sigma \in \mathrm{O}_{H^\perp}(V)$ such that $\sigma u = cu$ and $\sigma v = c^{-1}v$. Since $\mathrm{O}(V)$ is generated by transvections, of order 2, $\mathrm{O}(V)/\mathrm{O}'(V)$ is generated by elements of order 2, and is abelian, so it has exponent 2. Thus $\sigma^2 \in \mathrm{O}'(V)$. For any $y \in H^\perp$ we have $\rho_{u,y} \in \Omega = \mathrm{O}'(V)$, and so

$$\begin{aligned}\sigma^2\rho_{u,y}\sigma^{-2}\rho_{u,y}^{-1} &= \rho_{\sigma^2u,\sigma^2y}\rho_{u,-y} = \rho_{c^2u,y}\rho_{u,y}\\ &= \rho_{u,c^2y}\rho_{u,y} = \rho_{u,(c^2+1)y} \in \mathrm{O}''(V).\end{aligned}$$

Since $c^2 + 1 \neq 0$ every pseudotransvection is of the form $\rho_{u,(c^2+1)y}$ for appropriately chosen u and y. Thus $\mathrm{O}'(V) = \Omega \leq \mathrm{O}''(V)$, and hence $\mathrm{O}''(V) = \mathrm{O}'(V)$.

Suppose next that $|F| = 2$, so either $n \geq 6$ or else $n = 4$ and $m = 1$. Let (u, v) be a hyperbolic pair and set $H = \langle u, v\rangle$, so $V = H \oplus H^\perp$. By Proposition 12.8 $H^\perp$ has a basis $\{w_1, w_2, \ldots, w_{n-2}\}$ of nonsingular vectors, so $Q(w_i) = 1$, $1 \leq i \leq n-2$. Note that $B(w_1, w_i)$ cannot be 0 for all i, since $H^\perp$ is nondefective. Thus we may assume that $B(w_1, w_2) = 1$, and hence $Q(w_1 + w_2) = 1$. Set $W = \langle w_1, w_2 \rangle$, a nondefective subspace, so $V = W \bigoplus W^\perp$. Define $\sigma\colon W \to W$ via $\sigma w_1 = w_2$, $\sigma w_2 = w_1 + w_2$, so $\sigma \in \mathrm{GL}(W)$. Also $Q(\sigma w_1) = Q(w_2) = 1 = Q(w_1)$ and $Q(\sigma w_2) = 1 = Q(w_2)$, so $\sigma \in \mathrm{O}(W)$. Extend σ trivially, so that $\sigma \in \mathrm{O}_{W^\perp}(V) \leq \mathrm{O}(V)$. Note that $\sigma^3 = 1$, and so $\sigma = \sigma^{-2} \in \mathrm{O}'(V)$ as in the first part of the proof. Also

$$\rho_{u,w_1} = \rho_{u,w_1+w_2}\rho_{u,w_2} = \rho_{\sigma u,\sigma w_1}\rho_{u,w_2} = \sigma\rho_{u,w_2}\sigma^{-1}\rho_{u,w_2}^{-1} \in \mathrm{O}''(V).$$

It follows that $\rho_{u,w_i} \in \mathrm{O}''(V)$, $1 \leq i \leq n-2$, and hence that $\rho_{u,y} \in \mathrm{O}''(V)$, all $y \in H^\perp$, since y is a sum of w_i's. All transvections have that form, so $\mathrm{O}'(V) = \Omega \leq \mathrm{O}''(V)$, and $\mathrm{O}''(V) = \mathrm{O}'(V)$. □

Theorem 14.43. *Suppose that V is a singular nondefective quadratic space over a field F of characteristic 2, with $\dim V = n \geq 4$ and Witt index m. If $(n, m) \neq (4, 2)$ then $\mathrm{O}'(V)$ is a simple group.*

Proof. Iwasawa's Theorem 0.5. □

The identifications with simple groups of Chevalley type are as in Chapter 6. See page 58, and see Carter [**11**] for details.

Let us now consider the specific case of a nondefective quadratic space of dimension 4 and Witt index 2, i.e. $(n, m) = (4, 2)$.

Let W_1 and W_2 be hyperbolic planes, with respective hyperbolic pairs (x_1, y_1) and (x_2, y_2). Set $V = W_1 \otimes W_2$ and set $u_1 = x_1 \otimes x_2$, $v_1 = y_1 \otimes y_2$, $u_2 = y_1 \otimes x_2$, $v_2 = x_1 \otimes y_2$. If $v = a_1u_1 + b_1v_1 + a_2u_2 + b_2v_2 \in V$ define $Q(v) = a_1b_1 + a_2b_2$. Set $H_1 = \langle u_1, v_1 \rangle$ and $H_2 = \langle u_2, v_2 \rangle$.

Exercise

Show that Q is a quadratic form on V, that H_1 and H_2 are hyperbolic planes in V, and that $V = H_1 \bigoplus H_2$. Conclude that V is nondefective and that $(n, m) = (4, 2)$.

Take $\sigma_1 \in \mathrm{GL}(W_1)$, and say that σ_1 is represented by matrix $C = [c_{ij}]$ relative to the basis $\{x_1, y_1\}$. If $v \in V$ then a routine calculation shows that $Q((\sigma_1 \otimes 1)v) = (\det\sigma_1)Q(v)$. Similarly take $\sigma_2 \in \mathrm{GL}(W_2)$, say with σ_2 represented by $D = [d_{ij}]$ relative to the basis $\{x_2, y_2\}$, and verify that $Q((1 \otimes \sigma_2)v) = (\det\sigma_2)Q(v)$. Consequently if we choose σ_1, σ_2 so that $(\det\sigma_1)(\det\sigma_2) = 1$ then we have $Q((\sigma_1 \otimes \sigma_2)v) = Q(v)$, all $v \in V$, and therefore $\sigma_1 \otimes \sigma_2 \in \mathrm{O}(V)$.

It follows from the routine calculations mentioned above that, relative to the basis $\{u_1, u_2, v_1, v_2\}$, $\sigma_1 \otimes \sigma_2$ is represented by

$$\begin{bmatrix} c_{11}d_{11} & c_{12}d_{11} & c_{12}d_{12} & c_{11}d_{12} \\ c_{21}d_{11} & c_{22}d_{11} & c_{22}d_{12} & c_{21}d_{12} \\ c_{21}d_{21} & c_{22}d_{21} & c_{22}d_{22} & c_{21}d_{22} \\ c_{11}d_{21} & c_{12}d_{21} & c_{12}d_{22} & c_{11}d_{22} \end{bmatrix}.$$

From this we may compute the pseudodeterminant $\delta(\sigma_1 \otimes \sigma_2)$, either using $z_{(\sigma_1 \otimes \sigma_2)\mathcal{B}} = z_{\mathcal{B}} + \delta(\sigma_1 \otimes \sigma_2)$ or the formula for δ in the exercise on page 130. We find that $\delta(\sigma_1 \otimes \sigma_2) = 0$, so $\sigma_1 \otimes \sigma_2 \in \mathrm{SO}(V)$.

Write $\mathcal{G}$ to denote the subgroup

$$\{(\sigma_1, \sigma_2) \colon (\det \sigma_1)(\det \sigma_2) = 1\} \leq \mathrm{GL}(W_1) \times \mathrm{GL}(W_2).$$

We have a homomorphism $\varphi \colon \mathcal{G} \to \mathrm{SO}(V)$, given by $\varphi(\sigma_1, \sigma_2) = \sigma_1 \otimes \sigma_2$. If $\varphi(\sigma_1, \sigma_2) = 1_V$ then in terms of the representing matrix above we see that

$$c_{11}d_{11} = c_{22}d_{11} = c_{22}d_{22} = c_{11}d_{22} = 1$$

and

$$c_{12} = c_{21} = d_{12} = d_{21} = 0,$$

hence $c_{11} = c_{22}$, $d_{11} = d_{22}$, and therefore σ_1 and σ_2 are of the form $\sigma_1 = c1_{W_1}$, $\sigma_2 = c^{-1}1_{W_2}$ for some $c \in F^\star$. Thus

$$\operatorname{Im} \varphi \cong \mathcal{G}/\{(c1, c^{-1}1) \colon c \in F^\star\}$$

We aim next at showing that φ maps $\mathcal{G}$ onto $\mathrm{SO}(V)$.

Exercise

Given α, β, γ, $\delta \in F$ with $(\alpha, \beta, \gamma, \delta) \neq (0, 0, 0, 0)$, but with $\alpha\beta + \gamma\delta = 0$, show that there exist $(c_1, c_2) \neq (0, 0)$ and $(d_1, d_2) \neq (0, 0)$ such that

$$\alpha = c_1 d_1, \ \beta = c_2 d_2, \ \gamma = c_2 d_1, \ \delta = c_1 d_2.$$

We need a technical result.

Proposition 14.44. *In the setting above suppose that $\sigma \in \mathrm{SO}(V)$ and that $\sigma u_1 = u_1$. Then $\sigma\langle u_1, u_2\rangle = \langle u_1, u_2\rangle$.*

Proof. (*Sketch*) Say $\sigma u_2 = \alpha_1 u_1 + \alpha_2 u_2 + \beta_1 v_1 + \beta_2 v_2$. Then $B(u_1, \sigma u_2) = 0 = \beta_1$ and $0 = Q(\sigma u_2) = \alpha_2\beta_2$. We need $\beta_2 = 0$, so suppose $\beta_2 \neq 0$ and hence $\alpha_2 = 0$, $\sigma u_2 = \alpha_1 u_1 + \beta_2 v_2$. Note similarly that the coefficient of v_1 in σv_1 is 1, and the coefficient of v_2 in σv_2 is 0. With that information compute $z_{\sigma\mathcal{B}} = \sum_{i=1}^{2} (\sigma u_i)(\sigma v_i)$. Since $\delta(\sigma) = 0$ we have $z_{\sigma\mathcal{B}} = z_{\mathcal{B}}$. If λ_2 denotes the coefficient of u_2 in σv_2 one checks that $\beta_2\lambda_2$ is the coefficient both for $u_2 v_2$ and for 1 in $z_{\sigma\mathcal{B}}$. Comparing coefficients gives $\beta_2\lambda_2 = 1$ and $\beta_2\lambda_2 = 0$, a contradiction. □

Proposition 14.45. *The homomorphism φ maps $\mathcal{G}$ onto* $\mathrm{SO}(V)$.

Proof. Take $\sigma \in \mathrm{SO}(V)$ and say that $\sigma u_1 = a_{11}u_1 + a_{21}u_2 + b_{11}v_1 + b_{21}v_2$. Then $0 = Q(u_1) = Q(\sigma u_1) = a_{11}b_{11} + a_{21}b_{21}$, and by the exercise above there exist $(c_{11}, c_{21}) \neq (0,0)$ and $(d_{11}, d_{21}) \neq (0,0)$ with $a_{11} = c_{11}d_{11}$, $a_{21} = c_{21}d_{11}$, $b_{11} = c_{21}d_{21}$, and $b_{21} = c_{11}d_{21}$. Take $\tau_1 \in \mathrm{GL}(W_1)$ and $\tau_2 \in \mathrm{GL}(W_2)$ represented by the matrices $C = [c_{ij}]$ and $D = [d_{ij}]$, respectively, with second columns chosen so that $\det \tau_1 = \det \tau_2 = 1$, and hence $(\tau_1, \tau_2) \in \mathcal{G}$. Then $(\tau_1 \otimes \tau_2)u_1 = \sigma u_1$, and hence we may assume that $\sigma u_1 = u_1$ (by replacing σ by $(\tau_1^{-1} \otimes \tau_2^{-1})\sigma$). By Proposition 14.44 we have $\sigma u_2 \in \langle u_1, u_2 \rangle$, say $\sigma u_1 = \alpha u_1 + \beta u_2$ (note that $\beta \neq 0$). Next define $\eta_1 \in \mathrm{GL}(W_1)$ and $\eta_2 \in \mathrm{GL}(W_2)$ represented, respectively, by matrices

$$C_1 = \begin{bmatrix} 1 & \alpha \\ 0 & \beta \end{bmatrix} \quad \text{and} \quad C_2 = \begin{bmatrix} 1 & 0 \\ 0 & \beta^{-1} \end{bmatrix},$$

so $(\eta_1, \eta_2) \in \mathcal{G}$. Also $(\eta_1 \otimes \eta_2)u_1 = u_1 = \sigma u_1$ and

$$(\eta_1 \otimes \eta_2)u_2 = \eta_1 y_1 \otimes \eta_2 x_2 = (\alpha x_1 + \beta y_1) \otimes x_2 = \alpha u_1 + \beta u_2 = \sigma u_2.$$

Thus we may (and shall) assume further that $\sigma u_1 = u_1$ and $\sigma u_2 = u_2$.

Write $\sigma v_1 = \alpha_1 u_1 + \alpha_2 u_2 + \beta_1 v_1 + \beta_2 v_2$, and note that $0 = B(u_1, \sigma v_1) = \beta_2$, $1 = B(u_1, \sigma v_1) = \beta_1$, and $0 = Q(\sigma v_1) = \alpha_1\beta_1 = \alpha_1$, so $\sigma v_1 = \alpha_2 u_2 + v_1$, and similarly $\sigma v_2 = \lambda u_1 + v_2$ for some λ. Next use $\delta(\sigma) = 0$, and hence $\sum_i u_i v_i = \sum_i (\sigma u_i)(\sigma v_i)$ to see that $\lambda = \alpha_2$. But then an easy calculation shows that if we take $\sigma_1 = 1_{W_1}$ and define σ_2 via $\sigma_2 x_2 = x_2$, $\sigma_2 y_2 = \lambda x_2 + y_2$ then $(\sigma_1, \sigma_2) \in \mathcal{G}$ and $\varphi(\sigma_1, \sigma_2) = \sigma_1 \otimes \sigma_2 = \sigma$. □

The notation and settings above will be maintained in the next theorem and its proof.

Theorem 14.46. *Suppose that* char $F = 2$ *but* $|F| > 2$ *and let V be a quadratic space of dimension 4 and Witt index 2. Then*

$$\mathrm{O}'(V) \cong \mathrm{SL}(2, F) \times \mathrm{SL}(2, F).$$

Proof. Write $\mathcal{G}_1 = \mathrm{GL}(W_1) \times \mathrm{GL}(W_2) \geq \mathcal{G}$, and note that

$$\mathcal{G} \geq \mathrm{SL}(W_1) \times \mathrm{SL}(W_2).$$

Then

$$\begin{aligned} \mathcal{G}_1' &\geq \mathrm{GL}'(W_1) \times \mathrm{GL}'(W_2) = \mathrm{SL}(W_1) \times \mathrm{SL}(W_2) \\ &\geq \mathcal{G}' \geq \mathrm{SL}'(W_1) \times \mathrm{SL}'(W_2) = \mathrm{SL}(W_1) \times \mathrm{SL}(W_2) \end{aligned}$$

by Theorem 1.7 and the exercise on page 10, so $\mathcal{G}' = \mathrm{SL}'(W_1) \times \mathrm{SL}'(W_2)$. The map $\varphi \colon \mathcal{G} \to \mathrm{SO}(V)$ restricts to

$$\varphi' : \mathcal{G}' = \mathrm{SL}(W_1) \times \mathrm{SL}(W_2) \to \mathrm{SO}'(V) = \mathrm{O}'(V).$$

Clearly φ' is surjective, and

$$\operatorname{Ker}\varphi' = \operatorname{Ker}\varphi \cap [\mathrm{SL}(W_1) \times \mathrm{SL}(W_2)] = \{(a1, a^{-1}1) : a^2 = 1\} = 1.$$

□

Remark

The discussion above can be modified to show that if char $F \neq 2$ then $\mathrm{P}\Omega(V) \cong \mathrm{PSL}(2, F) \times \mathrm{PSL}(2, F)$, as was remarked on page 83.

If $|F| = 2$ then $\mathcal{G}_1 = \mathcal{G}$ and $\mathrm{GL}(W_i) = \mathrm{SL}(W_i)$, so φ maps $\mathrm{SL}(W_1) \times \mathrm{SL}(W_2)$ onto $\mathrm{SO}(V)$, and $\operatorname{Ker}\varphi = 1$, so $\mathrm{SO}(V) \cong \mathrm{SL}(2,2) \times \mathrm{SL}(2,2) \cong \mathrm{Sym}(3) \times \mathrm{Sym}(3)$, and so $\mathrm{SO}'(V) \cong \mathrm{Alt}(3) \times \mathrm{Alt}(3)$.

Remark

When $(n, m, |F|) = (4, 2, 2)$ then $\mathrm{O}(V)$ is isomorphic with a wreath product $\mathrm{Sym}(3) \wr \mathbb{Z}_2$.

Our final goal in this chapter is to determine the orders of the various groups we have been discussing in the case that the field F is finite.

If V is a hyperbolic plane then it follows easily from the exercise on page 137 that $\mathrm{SO}(V) \cong F^\star$ and that $\mathrm{O}(V)$ is a semidirect product , $\mathrm{O}(V) = \mathrm{SO}(V) \rtimes \langle\tau\rangle$ for any transvection τ, with $\tau\sigma\tau^{-1} = \sigma^{-1}$ for every $\sigma \in \mathrm{SO}(V)$. In particular, if $|F| = q$, finite, then $\mathrm{O}(V)$ is dihedral of order $2(q-1)$.

If $\dim V = 2$ but V is nonsingular (and nondefective) then by Proposition 12.7 there is a basis $\{v_1, v_2\}$ for V and an irreducible polynomial $x^2 + x + b$ in $F[x]$ such that $Q(a_1v_1 + a_2v_2) = a_1^2 + a_1a_2 + ba_2^2$, all $a_i \in F$. Let $K = F(c)$ be an extension field of F with $c^2 + c = b$. Much as in Exercise 2 on page 46 we have $\mathrm{SO}(V) \cong K^\star/F^\star$, and $\mathrm{O}(V)$ a semidirect product $\mathrm{SO}(V) \rtimes \langle\tau\rangle$, τ a transvection, with $\tau\sigma\tau^{-1} = \sigma^{-1}$, all $\sigma \in \mathrm{SO}(V)$. If $|F| = q$, finite, then $\mathrm{O}(V)$ is dihedral of order $2(q+1)$.

Note that although there are two inequivalent quadratic forms when F is perfect (in the above context, F finite, V nonsingular), by Proposition 13.14, only one group $\mathrm{O}(V) \cong D_{q+1}$ occurs.

Suppose now in general that $|F| = q$, a power of 2, and that V is a quadratic space over F, with $\dim V = n \geq 1$. By Theorem 12.9 (and its proof) V is one of three types:

Type 1.: $V = H_1 \oplus \cdots \oplus H_m \oplus \langle v\rangle$, $Q(v) = 1$, $n = 2m+1$;

Type 2.: $V = H_1 \oplus \cdots \oplus H_m$, $n = 2m$; or

Type 3.: $V = H_1 \oplus \cdots \oplus H_m \oplus W$, $\dim W = 2$, $n = 2m+2$;

where each H_i is a hyperbolic plane and W is nonsingular.

In all cases write $s_n = |\mathcal{C}| = |\{v \in V : Q(v) = 0\}|$ and write h_n for the number of distinct hyperbolic pairs in V.

Proposition 14.47. *(a) If V is of type 1 then*

$$s_n = q^{n-1}, \quad h_n = q^{n-2}(q^{n-1} - 1);$$

(b) if V is of type 2 then

$$s_n = q^{n-1} + q^m - q^{m-1}, \quad h_n = q^{n-2}(q^m - 1)(q^{m-1} + 1);$$

(c) if V is of type 3 then

$$s_n = q^{n-1} - q^{m+1} + q^m, \quad h_n = q^{n-2}(q^{m+1} + 1)(q^m - 1).$$

Proof. Induction on n. If $n = 1$ then V is of type 1 and clearly $s_1 = 1$ and $h_1 = 0$. Suppose that $n = 2$ and V is of type 3. Again $s_2 = 1$ and $h_2 = 0$. Suppose that $n = 2$ and V is of type 2, with hyperbolic pair (u, v). Then $Q(au + bv) = ab = 0$ if and only if $a = 0$ or $b = 0$, so

$$s_2 = q + q - 1 = 2q - 1 = q^{n-1} + q^m - q^{m-1}.$$

Thus there are $2q - 2$ choices for u in a hyperbolic pair. Given u choose v. If $(u, au + bv)$ is to be a hyperbolic pair we need $B(u, au + bv) = b = 1$, and also $Q(au + v) = a = 0$, so v is unique and $h_2 = 2(q - 1)$.

Suppose then that $n \geq 3$ and that the result holds for dimensions less than n. Write $V = H \oplus H^\perp$, $H = \langle u, v \rangle$, (u, v) a hyperbolic pair. Then $u^\perp = \langle u \rangle \oplus H^\perp$. The singular (or zero) vectors in V consist of those that are in $u^\perp$ and those that are not. First, those in $u^\perp$ are those of the form $au + w$, $w \in H^\perp$, with $0 = Q(au + w) = Q(w)$, so their number is $q \cdot s_{n-2}$. Next, those not in $u^\perp$ have the form $au + bv + w$, $w \in H^\perp$, with $b \neq 0$, but with

$$0 = Q(au + bv + w) = ab + Q(w).$$

There are $q - 1$ choices for b, then q^{n-2} choices for w, and then a is determined. Conclusion: $s_n = qs_{n-2} + (q - 1)q^{n-2}$. If n is odd then V is of type 1 and by induction $s_{n-2} = q^{n-3}$, so $s_n = q \cdot q^{n-3} + q^{n-1} - q^{n-2} = q^{n-1}$. If V is of type 2 then $H^\perp$ is of type 2, so $s_{n-2} = q^{n-3} + q^{m-1} - q^{m-2}$, and therefore

$$s_n = q(q^{n-3} + q^{m-1} - q^{m-2}) + q^{n-1} - q^{n-2} = q^{n-1} + q^m - q^{m-1}.$$

If V is of type 3 then $H^\perp$ is of type 3, and the argument is similar.

For hyperbolic pairs in V first choose u singular; there are $s_n - 1$ choices. Then choose some v such that (u, v) is hyperbolic and set $H = \langle u, v \rangle$. Next choose $v' = au + bv + w$, $w \in H^\perp$, such that (u, v') is also hyperbolic. That requires $B(u, v') = b = 1$ and $Q(v') = Q(au+v+w) = a+Q(w) = 0$. There are q^{n-2} choices for w and then a is determined, so $h_n = (s_n - 1)q^{n-2}$. Thus if n is odd then $h_n = (q^{n-1} - 1)q^{n-2}$. If V is of type 2 then

$$h_n = q^{n-2}(q^{2m-1} + q^m - q^{m-1} - 1) = q^{n-2}(q^m - 1)(q^{m-1} + 1).$$

Finally, if V is of type 3 then

$$h_n = q^{n-2}(q^{2m+1} - q^{m+1} + q^m - 1) = q^{n-2}(q^{m+1} + 1)(q^m - 1).$$

□

We are finally prepared to calculate the orders of the groups when F is finite. Note that the calculation is unnecessary when n is odd (type 1), as $\mathrm{O}(V)$ is symplectic in that case by Theorem 14.2, and the order was determined in Theorem 3.12.

Theorem 14.48. *Suppose that $|F| = q$ and V is a nondefective quadratic space over F. If V is of type 2 then*

$$|\,\mathrm{O}(V)| = 2q^{\frac{n(n-2)}{4}}(q^{\frac{n}{2}} - 1)\prod_{i=1}^{\frac{n-2}{2}}(q^{2i} - 1),$$

if of type 3 then

$$|\,\mathrm{O}(V)| = 2q^{\frac{n(n-2)}{4}}(q^{\frac{n}{2}} + 1)\prod_{i=1}^{\frac{n-2}{2}}(q^{2i} - 1).$$

Proof. Note that the formulas are correct when $n = 2$, giving orders of $2(q-1)$ and $2(q+1)$, respectively, so we may assume that $n \geq 4$. Thus hyperbolic pairs exist in V. Let H be a hyperbolic plane in V, so $V = H \oplus H^{\perp}$. By the Witt Extension Theorem (12.10) $\mathrm{O}(V)$ acts transitively on the set of hyperbolic pairs in V. If (u, v) is a hyperbolic pair in H then $\mathrm{Stab}_{\mathrm{O}(V)}((u, v)) = \mathrm{O}_H(V) \cong \mathrm{O}(H^{\perp})$, and $H^{\perp}$ is of the same type as V. Thus $|\,\mathrm{O}(V)| = h_n |\,\mathrm{O}(H^{\perp})|$, and we may assume $|\,\mathrm{O}(H^{\perp})|$ known by induction.

Type 2

$$\begin{aligned}
|\,\mathrm{O}(V)| &= q^{n-2}(q^{\frac{n}{2}} - 1)(q^{\frac{n-2}{2}} + 1)\cdot 2q^{\frac{(n-2)(n-4)}{4}}(q^{\frac{n-2}{2}} - 1)\prod_{i=1}^{\frac{n-4}{2}}(q^{2i} - 1)\\
&= 2q^{\frac{(n-2)(n-4)}{4}+n-2}(q^{n-2} - 1)(q^{\frac{n}{2}} - 1)\prod_{i=1}^{\frac{n-4}{2}}(q^{2i} - 1)\\
&= 2q^{\frac{n(n-2)}{4}}(q^{\frac{n}{2}} - 1)\prod_{i=1}^{\frac{n-2}{2}}(q^{2i} - 1).
\end{aligned}$$

Type 3

$$\begin{aligned}
|\,\mathrm{O}(V)| &= q^{n-2}(q^{\frac{n}{2}}+1)(q^{\frac{n-2}{2}}-1)\cdot 2q^{\frac{(n-2)(n-4)}{4}}(q^{\frac{n-2}{2}}+1)\prod_{i=1}^{\frac{n-4}{2}}(q^{2i}-1)\\
&= 2q^{\frac{n^2-6n+8+4n-8}{4}}(q^{\frac{n}{2}}+1)\prod_{i=1}^{\frac{n-4}{2}}(q^{2i}-1)\\
&= 2q^{\frac{n(n-2)}{4}}(q^{\frac{n}{2}}+1)\prod_{i=1}^{\frac{n-2}{2}}(q^{2i}-1).
\end{aligned}$$

□

Corollary 14.49. $|\mathrm{O}'(V)| = q^{\frac{n(n-2)}{4}}(q^{\frac{n}{2}}+\epsilon)\prod_{i=1}^{\frac{n-2}{2}}(q^{2i}-1)$, *where* $\epsilon = -1$ *if* V *is hyperbolic (type 2),* $\epsilon = 1$ *otherwise (type 3).*

Proof. We know that the pseudodeterminant δ maps $\mathrm{O}(V)$ onto $\mathbb{F}_2$, and so $|\,\mathrm{SO}(V)| = (1/2)|\,\mathrm{O}(V)|$. Since $F^\star = F^{\times 2}$ we have $\mathrm{SO}(V)$ contained in the kernel of the spinor norm θ by Proposition 14.18, and hence $\mathrm{O}'(V) = \mathrm{SO}(V)$ by Proposition 14.23. □

Note for example that if $(n,m,q) = (4,1,2)$ then $|\mathrm{O}'(4,2)| = 4\cdot 5\cdot 3 = 60$, so $\mathrm{O}'(4,2) \cong \mathrm{Alt}(5)$.

It is customary to write $\mathrm{O}_+(V)$ for $\mathrm{O}(V)$ if $m = n/2$, and $\mathrm{O}_-(V)$ if $m = (n-2)/2$.

It can be shown in general that $\mathrm{O}'_-(4,q) \cong \mathrm{PSL}(2,q^2)$ (see e.g. [**67**], page 201).

If $(n,m,q) = (6,3,2)$ then $|\mathrm{O}'_+(V)| = 20{,}160$. It can be shown that $\mathrm{O}'_+(6,2) \cong A_8 \cong \mathrm{PSL}(4,2)$. In fact $\mathrm{O}'_+(6,q) \cong \mathrm{PSL}(4,q)$ for all prime powers q ([**67**], page 190).

If $(n,m,q) = (6,2,2)$ then $|\mathrm{O}'_-(V)| = 25{,}920$, and $\mathrm{O}'_-(6,2) \cong \mathrm{PSU}(4,4)$, which is isomorphic to $\mathrm{PSp}(4,3)$ (see [**42**], page 220), which was mentioned in Chapter 3 as an index 2 subgroup of the symmetry group for the 27 lines on a general cubic surface. More generally $\mathrm{O}'_-(6,q) \cong \mathrm{PSU}(4,q)$ for all prime powers q ([**67**], page 198).

Chapter 15

Further Developments

As was mentioned in the preface the theory of groups of Chevalley type is based on the classification, due to Killing and Cartan, of simple Lie algebras over the complex field $\mathbb{C}$ (the book [**37**] by Humphreys is a good general reference for information about Lie algebras; also see [**7**]). The simple Lie algebras are classified by the *Coxeter-Dynkin diagrams* A_n for $n \geq 1$, B_n and C_n for $n \geq 2$, D_n for $n \geq 4$, and the five *exceptional* diagrams G_2, F_4, E_6, E_7, and E_8.

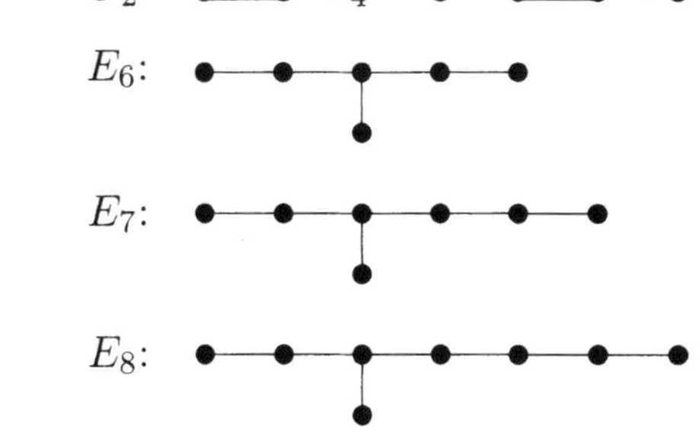

C. Chevalley [**13**] defined in 1955 a class of groups corresponding to a given simple Lie algebra, one for each field F, the elements of the groups being automorphisms of a resulting Lie algebra over F. He accomplished this by exhibiting a basis for the Lie algebra for which all structure constants are rational integers. Thus the $\mathbb{Z}$-module spanned by the basis is a Lie algebra over $\mathbb{Z}$, and tensoring with F yields a Lie algebra over F. The Chevalley group for F is then a group of automorphisms of the F-Lie algebra.

As indicated in earlier chapters the Chevalley groups include certain of the classical groups. Groups of type A_n are projective special linear groups, of type B_n are projective orthogonal groups in odd dimensions, of type C_n are projective symplectic groups, and of type D_n are hyperbolic projective orthogonal groups. The groups of type G_2 were discovered by L. E. Dickson [**19**] in 1905; Chevalley's work produced also simple groups of types E_6, E_7, and E_8, hitherto unknown.

Shortly after Chevalley's paper appeared R. Steinberg [**65**], J. Tits [**68**], and D. Hertzig [**35**] produced so-called *twisted* variations on the Chevalley groups, based on symmetries of the Coxeter-Dynkin diagrams. In particular, groups of type ${}^2A_n$ are the projective special unitary groups, and those of type ${}^2D_n$ are the remaining projective orthogonal groups in even dimensions. They are often called *Steinberg groups*, as are the twisted groups of type ${}^2E_6$ and ${}^3D_4$. The remaining twisted types are the *Suzuki groups* of type ${}^2B_2$ and the *Ree groups* of type ${}^2G_2$ and ${}^2F_4$.

Except for the alternating groups and the 26 sporadic groups every non-abelian finite simple group is a group of Chevalley type.

The book [**11**] by R. Carter is an excellent source of information about groups of Chevalley type.

J. Tits (see [**69**]) further unified the study of groups of Chevalley type, including the exceptional groups, by introducing the notion of groups with BN-pairs (sometimes called *Tits systems*). Such groups act in a natural way on geometrical objects called *buildings*, also introduced by Tits. The definitions are somewhat involved, so readers are referred to [**69**]; [**9**] and [**60**] are two more recent books on buildings.

Chevalley groups are special cases of *algebraic groups*. An algebraic group over a field F is a variety G defined over F for which the maps defining the group operations, viz. $\pi\colon G \times G \to G$ defined by $\pi(\sigma, \tau) = \sigma\tau$ and $\iota\colon G \to G$ defined by $\iota(\sigma) = \sigma^{-1}$, are morphisms of varieties.

For a very simple example consider $G = \mathrm{SL}(2, F)$. Its elements are precisely the set of roots of the polynomial

$$\det \begin{bmatrix} x_1 & x_2 \\ x_3 & x_4 \end{bmatrix} - 1 = x_1x_4 - x_2x_3 - 1$$

in $M_2(F)$, so it is an affine variety. The maps π and ι are polynomials in the matrix entries, hence are morphisms.

Several accounts of the theory and classification of algebraic groups are available; [**38**] and [**64**] are recommended as being particularly readable. The theory generally requires a reasonable background in algebraic geometry.

We move on to further developments in rather a different direction. It was mentioned in the preface that some of the spaces and groups in earlier chapters could have been defined over noncommutative division rings as well as over fields. In fact analogues of the classical groups have been studied at great length over a variety of classes of rings that are not fields, requiring appropriate generalizations of bilinear, sesquilinear, and quadratic forms, etc., on modules over the rings. All rings considered will be associative rings with 1.

Algebraic K-theory has its roots in geometric topology and algebraic geometry, but we will focus on connections with the classical groups. Its applications in general are wide-ranging, from algebraic topology to number theory to functional analysis, and much more. It is often viewed as the extension of linear algebra from vector spaces with fields as scalars to modules with rings as scalars. For some of the history of the subject see [**73**].

As we studied the classical groups above an important point in the development of each family was the application of Iwasawa's Theorem to conclude simplicity. When fields are replaced by more arbitrary rings simplicity tends to fail, and one of the important uses of K-theory is to help in determining the normal subgroup structure of classical groups over rings.

For each natural number n there is a functor K_n from the category of rings to the category of abelian groups. To begin: K_0 assigns to a ring R an additive abelian group $K_0(R)$ called the *Grothendieck group.* It is defined by generators and relations as follows. The generators are isomorphism classes $[P]$ of finitely generated projective (left) modules P over R, and for each pair $[P]$ and $[Q]$ of generators one relation, viz. $[P] + [Q] = [P \oplus Q]$. It follows easily that $K_0(R)$ is the set of all differences $[P] - [Q]$, where $[P]$ and $[Q]$ are any generators. In fact $K_0(R)$ is a commutative ring, based on $[P] \cdot [Q] = [P \otimes Q]$.

Since K_0 doesn't play a major role in the study of classical groups we move on to K_1, which requires somewhat more preparation.

View $\mathrm{GL}(n, R)$ as a subgroup of $\mathrm{GL}(n+1, R)$ as follows: identify σ in $\mathrm{GL}(n, R)$ with $\begin{bmatrix} \sigma & 0 \\ 0 & 1 \end{bmatrix} \in \mathrm{GL}(n+1, R)$. Define

$$\mathrm{GL}(R) = \bigcup_{n=1}^{\infty} \mathrm{GL}(n, R),$$

the *stable general linear group.*

An *elementary* matrix in $\mathrm{SL}(n, R)$ is a matrix that coincides with the identity matrix except for having a single nonzero entry off the diagonal. Write $E(n, R)$ for the subgroup of $\mathrm{GL}(n, R)$ generated by the elementary matrices, and with the convention above in force define

$$E(R) = \bigcup_{n=1}^{\infty} E(n, R),$$

the *stable elementary group*.

Then $E(R)$ is the derived group $\mathrm{GL}'(R)$ of $\mathrm{GL}(R)$ (for a proof see e.g. [**55**], page 25). Thus $\mathrm{GL}(R)/E(R)$ is an abelian group, and $K_1(R)$ is defined to be $\mathrm{GL}(R)/E(R)$.

For applications to normal subgroup structure the ideas above must be generalized.

Suppose that $\mathfrak{a}$ is an ideal in R. Then $\mathrm{GL}(\mathfrak{a})$ and $E(\mathfrak{a})$ are defined as above, $\mathrm{GL}(\mathfrak{a}) \lhd \mathrm{GL}(R)$, and $E(\mathfrak{a}) \lhd \mathrm{GL}(R)$ (see e.g. [**33**]). Define $K_1(\mathfrak{a})$ to be $\mathrm{GL}(\mathfrak{a})/E(\mathfrak{a})$.

A vector $(x_1, \ldots, x_k)$ in the free R-module R^k is called *unimodular* if there exist $t_1, \ldots, t_k \in R$ such that $\sum_{i=1}^{k} t_i x_i = 1$. If $1 \leq k \in \mathbb{N}$ the ring R satisfies the *stable range condition* $(S)_k$ if given a unimodular vector $(x_0, x_1, \ldots, x_k) \in R^{k+1}$ there exist $t_1, \ldots, t_k \in R$ such that $(x_1 + t_1 x_0, \ldots, x_k + t_k x_0)$ is unimodular in R^k. If R satisfies $(S)_k$ it also satisfies $(S)_{k+1}$, and hence also all $(S)_m$ for $m \geq k$. If R satisfies $(S)_m$ for some m then the *stable rank* of R is the least k for which R satisfies $(S)_k$.

If $\mathfrak{a}$ is an ideal in R define $\mathrm{GL}^{\sim}(n, R)$ to be the kernel of the composite map

$$\mathrm{GL}(n, R) \to \mathrm{GL}(n, R/\mathfrak{a}) \to \mathrm{PGL}(n, R/\mathfrak{a}),$$

the two maps being the obvious ones.

The main theorem regarding normal subgroup structure of $\mathrm{GL}(n, R)$ follows. It is due to H. Bass, with improvements by J. Wilson, I. Golubchik, and L. Vaserstein. See [**31**] or [**33**] for references. The proof relies on the K_1 functor.

Theorem 15.1. *Suppose that n is greater than both 2 and the stable rank of R, and that H is a subgroup of $\mathrm{GL}(n, R)$ that is normalized by $E(n, R)$. Then there is a unique ideal $\mathfrak{a}$ in R such that $E(n, \mathfrak{a}) \leq H \leq \mathrm{GL}^{\sim}(n, R)$. Conversely, if $E(n, \mathfrak{a}) \leq H \leq \mathrm{GL}^{\sim}(n, R)$ for some ideal $\mathfrak{a}$ then H is normalized by $E(n, R)$.*

To give some idea of consequences of the theorem we cite a few examples of stable ranks for some familiar classes of rings (see [**33**] for more). Local rings and semi-simple Artinian rings have stable rank 1. Dedekind domains have stable rank ≤ 2. If D is a Dedekind domain then the polynomial ring $R = D[x_1, \ldots, x_k]$ has stable rank $\le k + 2$.

For another application of K_1 we appeal to an interaction between the classical groups and number theory (see e.g. [**62**]). If $\mathfrak{a}$ is a proper nontrivial ideal in $\mathbb{Z}$ then $\mathrm{SL}(n, \mathfrak{a})$ is called a *congruence* subgroup of $\mathrm{SL}(n, \mathbb{Z})$. It is the kernel of the natural homomorphism from $\mathrm{SL}(n, \mathbb{Z})$ onto $\mathrm{SL}(n, \mathbb{Z}/\mathfrak{a})$, a finite group, hence has finite index in $\mathrm{SL}(n, \mathbb{Z})$.

The *congruence subgroup problem* for $\mathrm{SL}(n, \mathbb{Z})$ asks for a converse. Does a subgroup of finite index in $\mathrm{SL}(n, \mathbb{Z})$ necessarily contain a congruence subgroup? The answer has been known to be "No" for the *modular group* $\mathrm{SL}(2, \mathbb{Z})$ for over a century (see Chapter VIII of [**56**] for examples). Surprisingly, perhaps, the answer is "Yes" for $n \ge 3$. Independent proofs, depending on the functor K_1, due to J. Mennicke and to H. Bass, M. Lazard, and J-P. Serre appeared in 1964/65. Related results are known for larger classes of rings, e.g. for rings of algebraic integers in global fields; see [**33**] and [**39**] for details.

The K_1 functor also plays a role in determining generating sets for classical groups.

The K_2 functor, whose development we omit, is also important for the study of classical groups over rings. The main applications are the determination of presentations for the groups and study of the isomorphisms between them. Again the main first reference is [**33**].

Definitions of the "higher" K_n functors, for $n \ge 3$, have basically been agreed upon, and the theory is still very much under development. See [**61**] for some of the details and also [**73**] for some of the history.

Strictly speaking, the K-theory sketched above should have been called *linear* K-theory. The definitions and theorems mentioned applied to the linear classical groups, i.e. GL, SL, and PSL. There is a very general notion of *form ring* which admits a broad generalization of the notion of quadratic form which can be specialized to all of the types of quadratic forms that have been discussed in earlier chapters, now even over rings. Their isometry groups specialize to classical groups of the various types (symplectic, orthogonal, unitary – even linear) over rings. There is a resulting *unitary* K-theory, with applications similar to those sketched above. See [**33**] for details and further references.

There has been considerable study of the subgroup structure of classical groups, beginning with Galois. Dickson (see [**18**] or [**41**]) classified all subgroups of $\mathrm{PSL}(2, q)$. In recent years there has been a flurry of studies of subgroups of simple groups, concentrating first on maximal subgroups.

For the classical finite simple groups the seminal paper was written by Aschbacher [**4**] in 1984. For a classical group G Aschbacher defined a family $\mathcal{C}(G)$ of subgroups, typically determined by natural geometric conditions, and he showed that almost every subgroup of G is contained in a subgroup in $\mathcal{C}(G)$.

In the book ([**47**], 1990) of Kleidman and Liebeck they carried Aschbacher's ideas considerably further. In particular, building on the work of Aschbacher and others, they (quoting from [**47**]) determine

(I): the group-theoretic structure of each member of $\mathcal{C}(G)$,

(II): the conjugacy among the members of $\mathcal{C}(G)$,

(III): precisely which members of $\mathcal{C}(G)$ are maximal in G and which are not — and, for non-maximal members H of $\mathcal{C}(G)$, we determine the maximal overgroups of H in G.

In ([**53**], 1998) Liebeck and Seitz shifted the emphasis from subgroups of finite classical groups to subgroups of classical groups over algebraically closed fields. The shift in emphasis resulted in simplifications of the arguments in [**4**], and in particular to a somewhat simpler re-definition of $\mathcal{C}(G)$, and to a classification theorem for maximal (or almost maximal) subgroups. Since the field is algebraically closed the subgroups are, of course, infinite. Nevertheless, they were able to obtain Aschbacher's original results for finite groups as a consequence by applying standard techniques involving fixed point sets under Frobenius maps.

Let us go into somewhat more detail, basically following [**53**]. It is convenient to deal with simple and not-quite-simple groups in several ways. A group G is *almost simple* if there is a simple group G_0 with $G_0 \leq G \leq \mathrm{Aut}(G_0)$. Say that G is *perfect* if $G' = G$, and that G is *quasisimple* if G is perfect and G modulo its center is simple.

Usage is slightly different for algebraic groups. A *simple algebraic group* is an algebraic group G that has no closed connected normal subgroups other than G and 1. It is in fact true (see e.g. [**38**]) that if G is a simple algebraic group then G has a finite center Z and G/Z is a simple group in the conventional sense.

Suppose then that F is an algebraically closed field, V is a vector space of dimension n over F, and that B is one of the following bilinear forms on V:

1. B is trivial, i.e. $B(u, v) \equiv 0$,
2. B is a nondegenerate alternating form, or
3. B is a nondegenerate orthogonal form.

Write $\mathrm{I}(B)$ for the group of B-isometries of V. Thus $\mathrm{I}(V)$ is either $\mathrm{GL}(V)$, $\mathrm{Sp}(V)$, or $\mathrm{O}(V)$ in the cases under consideration. Write $\mathrm{Cl}(V)$ for the derived group $\mathrm{I}'(V)$, either $\mathrm{SL}(V)$, $\mathrm{Sp}(V)$, or $\mathrm{SO}(V)$. Note that in almost all cases $\mathrm{Cl}(V) = \mathrm{I}(V) \cap \mathrm{SL}(V)$. Except for SL_1, SO_1, SO_2, and SO_4, each $\mathrm{Cl}(V)$ is a simple algebraic group.

The class $\mathcal{C}(G)$ of subgroups M of a simple classical algebraic group $G = \mathrm{Cl}(V)$ is the union of six subclasses $\mathcal{C}_1, \ldots, \mathcal{C}_6$, which we proceed to describe.

- $\mathcal{C}_1$: *Subspace stabilizers* $M = G_U$, where U is a proper nonzero subspace of V, which is nondegenerate or totally singular, or, if char $F = 2$ and $G = \mathrm{SO}(V)$ then U is nonsingular of dimension 1.
- $\mathcal{C}_2$: *Stabilizers of orthogonal decompositions.* In this case $V = \bigoplus_1^t V_i$, with $t > 1$ and $V_1, \ldots, V_t$ mutually orthogonal and isometric, and $M = G_{V_1,\ldots,V_t}$, which is isomorphic with a wreath product $(\mathrm{I}(V_1) \wr \mathrm{Sym}_t) \cap G$.
- $\mathcal{C}_3$: *Stabilizers of totally singular decompositions.* Here $V = U \otimes W$, where U, W are maximal totally singular subspaces, $G = \mathrm{Sp}(V)$ or $\mathrm{SO}(V)$, and $M = G_{U,W}$. If dim $V = 2m$, even, then M is an extension of GL_m by a cyclic group of order 2 or, if $G = \mathrm{SO}(V)$ and m is odd, then $M = \mathrm{GL}_m$.
- $\mathcal{C}_4$: *Tensor product subgroups.* There are two cases.
  1. $V = V_1 \otimes V_2$, with dim $V_i > 1$, and $M = N_G(\mathrm{Cl}(V_1) \circ \mathrm{Cl}(V_2))$. Then M acts naturally on $V_1 \otimes V_2$ and is denoted $\mathrm{Cl}(V_1) \otimes \mathrm{Cl}(V_2)$. The actual tensor product subgroups $\mathrm{Cl}(V_1) \otimes \mathrm{Cl}(V_2)$ that occur are
     - $\mathrm{SL} \otimes \mathrm{SL} < \mathrm{SL}$,
     - $\mathrm{Sp} \otimes \mathrm{Sp} < \mathrm{SO}$,

     and, if $p \neq 2$,
     - $\mathrm{Sp} \otimes \mathrm{SO} < \mathrm{Sp}$, and
     - $\mathrm{SO} \otimes \mathrm{SO} < \mathrm{SO}$.
  2. $V = \bigotimes_1^t V_i$, $t > 1$, the subspaces V_i mutually isometric, and $M = N_G(\prod \mathrm{Cl}(V_i))$. As in (1) of $\mathcal{C}_4$ M is denoted by $\bigotimes_1^t \mathrm{Cl}(V_i)$. The tensor product subgroups $\bigotimes_1^t \mathrm{Cl}(V_i)$ that occur are
     - $\bigotimes_1^t \mathrm{SL} < \mathrm{SL}$;
     - $\bigotimes_1^t \mathrm{Sp} < \mathrm{Sp}$ if t is odd and $p \neq 2$;
     - $\bigotimes_1^t \mathrm{Sp} < \mathrm{SO}$ if t is even or $p = 2$; and
     - $\bigotimes_1^t \mathrm{SO} < \mathrm{SO}$ if $p \neq 2$ and dim V_i is neither 2 nor 4.
- $\mathcal{C}_5$: *Finite local subgroups.* Suppose that r is prime, $r \neq p$, and let R be an extraspecial r-group of order r^{1+2m} and exponent $r \cdot (2, r)$

or, if $r = 2$, a central product of such a group with a cyclic group of order 4 (see e.g. [**22**] for the definition and properties of extraspecial groups). The group R embeds as a subgroup in $G = \mathrm{Cl}(V)$, where $\dim V = n = r^m$, V being an irreducible representation module over $F = \overline{\mathbb{F}}_p$ for R. There are some technical restrictions regarding r and the center $Z(R)$; see [**53**] for details. In this case $M = N_G(R)$, often referred to as the "normalizer of a cross-characteristic embedded extraspecial subgroup".

- $\mathcal{C}_6$: *Classical subgroups.* $M = N_G(\mathrm{Sp}(V))$ or $N_G(\mathrm{SO}(V))$ in $G = \mathrm{SL}(V)$, and $M = N_G(\mathrm{SO}(V))$ in $G = \mathrm{Sp}(V)$ with $p = 2$.

We may now define the class $\mathcal{C}(G)$ to be $\cup_{i=1}^{6}\mathcal{C}_i$.

If $H \leq \mathrm{GL}(V)$ and $V = V_1 \otimes V_2$, with $\dim V_i > 1$, say that H is *tensor-decomposable* if $H \leq \mathrm{GL}(V_1) \otimes \mathrm{GL}(V_2)$ (as in the description of subgroups in class $\mathcal{C}_4$ above). If not then H is *tensor-indecomposable.*

We may now state the main result of [**53**].

Theorem 15.2 (Liebeck-Seitz, 1998). *Suppose that $G = \mathrm{Cl}(V)$ is a classical simple algebraic group, and that H is a closed subgroup of G. Then one of the following holds:*

1. *H is a subgroup of a member of $\mathcal{C}(G)$;*
2. *modulo scalars H is almost simple, and $E(H)$ (the unique quasisimple normal subgroup of H) acts irreducibly on V and, if $G = \mathrm{SL}(V)$, $E(H)$ fixes no nondegenerate bilinear form on V; further, if H is infinite then $E(H)$ is tensor-indecomposable on V.*

The proof depends on linear algebra and basic results about algebraic groups.

In order to state Liebeck-Seitz's version of Aschbacher's theorem for subgroups of finite simple classical groups more concepts and terminology will be required.

Let K be an algebraically closed field of characteristic $p > 0$, and suppose that $G \leq \mathrm{GL}(n, K)$, a matrix group. Say that $q = p^f$ and define a map $\sigma_q\colon G \to \mathrm{GL}(n, K)$ via $\sigma_q(a_{ij}) = (a_{ij}^q)$, evidently a homomorphism. If $\sigma_q(G) \leq G$ then σ_q is called a *standard Frobenius map* on G. More generally a homomorphism $\sigma\colon G \to G$ is called a *Frobenius map* on G if some power of σ is a standard Frobenius map.

If σ is a Frobenius map on G write G_σ for the fixed point group of σ, i.e. $G_\sigma = \{\tau \in G\colon \sigma(\tau) = \tau\}$. It is a fact that finite classical groups arise as fixed point groups for Frobenius maps. For example (from [**47**]), if $G = \mathrm{SL}(n, K)$ set $\sigma_1 = \sigma_q$ and say that $\sigma_2\colon (a_{ij}) \mapsto ((a_{ij}^q)^t)^{-1}$, so that $\sigma_2^2 = \sigma_{q^2}$. Then $G_{\sigma_1} = \mathrm{SL}(n, q)$ and $G_{\sigma_2} = \mathrm{SU}(n, q)$.

The restriction of a Frobenius map σ to the fixed point group G_σ is an automorphism of G_σ, called a *Frobenius automorphism.*

Another group of automorphisms arises from symmetries of Coxeter-Dynkin diagrams (see page 151). These are called *graph automorphisms.* Chevalley groups of type D_4 in particular admit "triality" automorphisms due to 3-fold rotational symmetry of the diagram. See Carter [**11**] for details.

Liebeck and Seitz describe the analogs $\mathcal{C}_i^\sigma$ for G_σ of the $\mathcal{C}_i$, $1 \leq i \leq 6$. They also describe further classes of subgroups of G_σ, viz. $\mathcal{C}_1'^\sigma$, $\mathcal{C}_2'^\sigma$, $\mathcal{C}_4'^\sigma$, $\mathcal{C}_4''^\sigma$, $\mathcal{C}_7^\sigma$, and $\mathcal{C}_7'^\sigma$. For example, class $\mathcal{C}_2'^\sigma$ has subgroups that result from field extensions of $\mathbb{F}_q$, class $\mathcal{C}_7^\sigma$ has subgroups that result from restricting to a subfield of $\mathbb{F}_q$, etc. See [**53**] for the technical details.

To state the final theorem let $K = \overline{\mathbb{F}}_p$, an algebraic closure of $\mathbb{F}_p$. Let $q = p^f > 1$ and let σ be a Frobenius automorphism of G with fixed point group G_σ a finite classical group over $\mathbb{F}_q$. It is shown in [**47**] that either G_σ' is quasisimple, or $G_\sigma' = \Omega_+(4, q)$, or G_σ' is solvable for a few small values of n and q.

Finally, the Liebeck-Seitz version of Aschbacher's theorem.

Theorem 15.3. *Let G_σ be a finite classical group over $\mathbb{F}_q$ as above, with G_σ' quasisimple. Suppose that H is a subgroup of G_σ which is invariant under a group S of automorphisms of G_σ. If $G_\sigma' = \mathrm{Sp}(4, 2^e)$ or $\Omega_+(8, q)$ assume that S does not contain an element involving a graph or triality automorphism, respectively. Then one of the following holds:*

1. *the subgroup H lies in an S-invariant member of one of the classes $\mathcal{C}_1^\sigma, \mathcal{C}_1'^\sigma, \ldots, \mathcal{C}_7'^\sigma$ listed above; or*
2. *modulo scalars H is almost simple, and the action of $E(H)$ on the natural module for G_σ is absolutely irreducible, and can be realized over no proper subfield of $\mathbb{F}_q$ (or of $\mathbb{F}_{q^2}$ if G_σ is unitary); moreover, if $G_\sigma = \mathrm{SL}(n, q)$, then $E(H)$ fixes no nondegenerate bilinear or unitary form on the natural module.*

Bibliography

[1] E. Artin, *The Orders of the Linear Groups,* Comm. Pure Appl. Math. **8** (1955), 355-366.

[2] E. Artin, *The Orders of the Classical Simple Groups,* Comm. Pure Appl. Math. **8** (1955), 455-472.

[3] E. Artin, *Geometric Algebra,* Interscience, New York, 1957.

[4] M. Aschbacher, *On the Maximal Subgroups of the Finite Classical Groups,* Invent. Math **76** (1984), 469-514.

[5] M. Aschbacher, *Finite Group Theory,* Cambridge University Press, Cambridge, 1986.

[6] H. Bass, *Algebraic K-theory,* W.A. Benjamin, Inc., Reading, MA, 1968.

[7] N. Bourbaki, *Groupes et Algèbras de Lie, 4, 5, 6,* Hermann, Paris, 1968.

[8] R. Brauer and H. Weyl, *Spinors in n Dimensions,* Amer. J. Math. **57** (1935), 425-449.

[9] K. S. Brown, *Buildings,* Springer-Verlag, New York, 1989.

[10] P. J. Cameron, *Permutation Groups,* Cambridge University Press, Cambridge, 1999.

[11] R. W. Carter, *Simple Groups of Lie Type,* John Wiley and Sons, London, 1972.

[12] C. Chevalley, *The Algebraic Theory of Spinors,* Columbia University Press, New York, 1954.

[13] C. Chevalley, *Sur Certain Groupes Simples,* Tohoku Math. J. (2) **7** (1955), 14-66.

[14] W. K. Clifford, *Applications of Grassmann's Extensive Algebra,* Amer. J. Math. **1** (1878), 350-358.

[15] J. H. Conway, R. T. Curtis, S. P. Norton, R. A. Parker, and R. A. Wilson, *Atlas of Finite Groups,* Clarendon Press, Oxford, 1985.

[16] C. W. Curtis, *The Classical Groups as a Source of Algebraic Problems,* Amer. Math. Monthly **74**, No. 1 (1967), 80-91.

[17] C. W. Curtis, *Linear Algebra,* Springer-Verlag, New York, 1984.

[18] L. E. Dickson, *Linear Groups with an Exposition of the Galois Field Theory,* Teubner, Leipzig , 1901; Dover, New York, 1958.

[19] L. E. Dickson, *A New System of Simple Groups,* Math. Ann. **60** (1905), 137-150.

[20] J. Dieudonné, *Sur les Groupes Classiques,* Actualités Sci. Ind., No. 1040, Hermann, Paris, 1948.

[21] J. Dieudonné, *La Géométrie des Groupes Classiques,* Berlin, 1955.

[22] L. Dornhoff, *Group Representation Theory,* Marcel Dekker, Inc., New York, 1971.

[23] P. Du Val, *Homographies, Quaternions and Rotations,* Oxford University Press, New York, 1964.

[24] N. Elkies, The Klein Quartic in Number Theory, in *The Eightfold Way,* S. Levy, Ed., MSI Pubs. 35, Cambridge U. Press, Cambridge, 1999.

[25] P. Garrett, *Buildings and Classical Groups,* Chapman and Hall, London, 1997.

[26] M. Goossens, F. Mittelbach, and A. Samarin, *The LaTeX Companion,* Addison-Wesley, New York, 1994.

[27] L. C. Grove, *Dickson's Pseudodeterminant Without Matrices,* Proc. Amer. Math. Soc. **86** (1982), 695-696.

[28] L. C. Grove, *Algebra,* Academic Press, New York, 1983.

[29] L. C. Grove and C. T. Benson, *Finite Reflection Groups,* 2nd ed., Springer-Verlag, New York, 1985.

[30] L. C. Grove, *Groups and Characters,* John Wiley and Sons, New York, 1997.

[31] A. J. Hahn, *Algebraic K-theory, Morita Theory, and the Classical Groups,* in Lecture Notes in Mathematics **1185**, Springer-Verlag, 1986.

[32] A. J. Hahn, *Quadratic Algebras, Clifford Algebras, and Arithmetic Witt Groups,* Springer-Verlag, New York, 1994.

[33] A. J. Hahn and O. T. O'Meara, *The Classical Groups and K-Theory,* Springer-Verlag, Berlin, 1989.

[34] P. R. Halmos, *Finite-Dimensional Vector Spaces,* Springer-Verlag, New York, 1974.

[35] D. Hertzig, *Forms of Algebraic Groups,* Proc. Amer. Math. Soc. **12** (1961), 657-660.

[36] D. G. Higman, *Finite Permutation Groups of Rank 3,* Math. Z. **86** (1964), 145-156.

[37] J. E. Humphreys, *Introduction to Lie Algebras and Representation Theory,* Springer-Verlag, New York, 1972.

[38] J. E. Humphreys, *Linear Algebraic Groups,* Springer-Verlag, New York, 1975.

[39] J. E. Humphreys, *Arithmetic Groups,* Lecture Notes in Mathematics **789**, Springer-Verlag, Berlin, 1980.

[40] J. E. Humphreys, *Reflection Groups and Coxeter Groups,* Cambridge University Press, Cambridge, 1990.

[41] B. Huppert, *Endliche Gruppen I,* Springer-Verlag, Berlin, 1967.

[42] B. Huppert, *Geometric Algebra,* U. of Illinois Chicago Lecture Notes, 1968/1969.

[43] K. Iwasawa, *Über die Einfachheit der Speziellen Projektiven Gruppen,* Proc. Imp. Acad. Tokyo **17** (1941), 57-59.

[44] N. Jacobson, *Basic Algebra I,* W. H. Freeman and Company, San Francisco, 1974.

[45] N. Jacobson, *Basic Algebra II,* W. H. Freeman and Company, San Francisco, 1980

[46] C. Jordan, *Traité des Substitutiones et des Équations Algébriques,* Gauthier-Villars, Paris, 1870.

[47] P. Kleidman and M. Liebeck, *The Subgroup Structure of the Finite Classical Groups,* Cambridge University Press, Cambridge, 1990.

[48] B. Kleines, *Die Einfachheit der Klassischen Gruppen,* Diplomarbeit, RWTH Aachen, 1989.

[49] D. Knuth, *The TEXbook,* Addison-Wesley, Reading, Mass., 1994.

[50] T.Y. Lam, *Algebraic Theory of Quadratic Forms,* W.A. Benjamin, Inc., Reading, MA, 1973.

[51] L. Lamport, *LATEX User's Guide and Reference Manual,* Addison-Wesley, Reading, Mass, 1994.

[52] M. Liebeck, *Introduction to the Subgroup Structure of Algebraic Groups,* in *Representations of Reductive Groups,* eds. R. Carter and M. Geck, Cambridge University Press, Cambridge, 1998.

[53] M. Liebeck and G. Seitz, *On the Subgroup Structure of Classical Groups,* Invent. Math. **134** (1998), 427-453.

[54] P. Lounesto, *Clifford Algebras and Spinors,* Cambridge University Press, Cambridge, 1997.

[55] J. Milnor, *Introduction to Algebraic K-theory,* Princeton University Press, Princeton, 1971.

[56] M. Newman, *Integral Matrices,* Academic Press, New York, 1972.

[57] D. S. Passman, *Permutation Groups,* Benjamin, New York, 1968.

[58] R. S. Pierce, *Associative Algebras,* Springer-Verlag, New York, 1982.

[59] I. R. Porteous, *Clifford Algebras and the Classical Groups,* Cambridge University Press, Cambridge, 1995.

[60] M. Ronan, *Lectures on Buildings,* Academic Press, Boston, 1989.

[61] J. Rosenberg, *Algebraic K-Theory and its Applications,* Springer-Verlag, New York, 1994.

[62] B. Schoeneberg, *Elliptic Modular Functions,* Springer-Verlag, Berlin, 1974.

[63] C.-L. Siegel, *Über die Analytische Theory der Quadratischen Formen II,* Ann. Math. **36** (1935), 230-263.

[64] T. Springer, *Linear Algebraic Groups, 2nd ed.,* Birkhäuser, Boston, 1998.

[65] R. Steinberg, *Variations on a Theme of Chevalley,* Pacific J. Math. **9** (1959), 875-891.

[66] R. Steinberg, *Lectures on Chevalley Groups,* Yale University Lecture Notes, 1968.

[67] D. E. Taylor, *The Geometry of the Classical Groups,* Heldermann Verlag, Berlin, 1992.

[68] J. Tits, *Sur la Trialité et Certain groupes qui s'en Déduisent,* Inst. Haute Études Sci. Publ. Math. **2** (1959), 14-60.

[69] J. Tits, *Buildings of Spherical Type and Finite BN-pairs,* Lecture Notes in Mathematics **386**, Springer-Verlag, Berlin, 1974.

[70] B. L. van der Waerden, *Gruppen von Linearen Transformationen,* Springer, Berlin, 1935; Chelsea, New York, 1948.

[71] B. L. van der Waerden, *A History of Algebra,* Springer-Verlag, Berlin, 1985.

[72] G. E. Wall, *The Structure of a Unitary Factor Group,* Publ. Math. IHES, No. 1, Paris, 1959.

[73] C. A. Weibel, *The Development of Algebraic K-theory Before 1980,* available from http://www.rutgers.math.edu/~weibel.

[74] H. Weyl, *The Classical Groups,* Princeton University Press, Princeton, 1946.

[75] D. Wick, *The Infamous Boundary,* Copernicus, New York, 1996.

List of Notation

Index